Containment systems

A design guide

Containment systems

A design guide

Edited by Nigel Hirst, Mike Brocklebank and Martyn Ryder

G|P
P|❦ **Gulf Professional Publishing**
an imprint of Elsevier Science

Published by
Institution of Chemical Engineers (IChemE)
Davis Building
165–189 Railway Terrace
Rugby, Warwickshire CV21 3HQ, UK
IChemE is a Registered Charity

Published exclusively in the USA and Canada by
Gulf Professional Publishing
An imprint of Elsevier Science
225 Wildwood Avenue
Woburn, Massachusetts 01801-2041
USA

© 2002 Institution of Chemical Engineers

ISBN 0 7506 7612 4   Gulf Professional Publishing

Typeset by Techset Composition Limited, Salisbury, UK

Printed by Bell & Bain Limited, Glasgow, UK

Preface

When the IChemE Pharma Subject Group was approached by a major contractor for information on current best practice in containment design, it became apparent that there was no single volume which collected together current thinking on the issue. During the writing of this book, the interest in containment has increased tremendously, as evidenced by the number of papers and conferences on the subject.

The aim of this book is to provide a practical design guide for anyone involved in the handling of toxic solids and liquids, with the main focus being the containment of toxic dusts during transfer operations. Often these materials are part of pharmaceutical manufacture, but other industries with challenges which are just as significant, such as fine chemicals, dyestuffs and agrochemicals, have been considered. Nuclear and biological containment are specialist areas outside the remit of this guide.

During preparation of the book, it became obvious that the many words used to describe aspects of containment were not uniquely defined. To avoid confusion we have provided definitions of key words as they are used in this guide, and have avoided, wherever possible, the use of general terms.

The structured approach to the design of containment systems — the 'Containment Strategy' — was developed over a period of time, and we are indebted to Steve Maidment of the Health and Safety Executive for access to the thinking of the Health and Safety Commission's Advisory Committee on the Safety of Chemicals. In the interests of compatibility, we have adopted the same concentration ranges for the Hazard Groups, extending them where necessary into areas of application in the industries this guide is concerned with.

The assessment of equipment performance has been supported by information supplied by many manufacturers and users. Examination has shown that the 'real-life' performance of containment systems is dependent on a variety of factors such as maintenance and operational standards, and not just the theoretical capability of the containment equipment installed.

As the toxicity of many new drugs increases, the need for containment systems grows. Recent figures indicate that the percentage of compounds handled in the pharmaceutical industry which are considered 'potent' increased from 5% in 1990 to 30% in 2000. The range of equipment which is available is constantly increasing, but it is hoped that the methodology outlined in this guide will become a firm basis on which to design any containment solution.

Nigel Hirst

Acknowledgements

Members of the IChemE working party on Containment:

Nigel Hirst (Chairman)   Haden Freeman

Mike Brocklebank   Foster Wheeler
John Challenger   Shepherd Process
John Dunn   GlaxoSmithKline
Kathleen Hoskins   Respirex
Steve Lewis   GlaxoSmithKline
Ewart Richardson   Isolation Solutions
Martyn Ryder   Extract Technology
David Smith   Tanvec
Steve Taylor   GlaxoSmithKline

In addition, the editors extend their thanks to Steve Maidment of the Health and Safety Executive, who provided important information on the approach taken by the ACTS committee, and to Gerald Cerulli of Foster Wheeler USA, who provided the text on US standards.

Special thanks are due to our technical editor, Peter Burke, whose patience, diligence and attention to detail are much appreciated.

The following companies kindly provided diagrams and photographs:
American Conference of Governmental Industrial Hygienists
Bosch Packaging
Drum Vent Systems
Extract Technology Ltd
Flomat Bagfilla International Ltd
GEA Buck Valve GmbH
Guyson International Ltd
MATCON
Process Containment Ltd
Trox Bros

Contents

Introduction

1

Purpose

The purpose of this chapter is to give an overview of the objectives and contents of this guide to the design of containment systems. Basic terms relevant to an understanding of the concepts of containment are explained and a summary is given of the contents of the remaining chapters of the guide.

Contents

What is containment?

Manufacturing industries, particularly pharmaceutical companies, are now using and producing materials that in very small quantities can have a significant adverse effect on people or the environment. It is therefore necessary to protect the population at large and in particular the employees working in those industries from the effects of these materials. It may also be necessary to protect products from contaminants in the workplace in order to ensure their uniformity and purity. These dual requirements are illustrated in Figure 1.1 (page 2).

The terms 'containment' and 'isolation' are often used interchangeably to describe equipment or systems that prevent release of hazardous materials. For the purposes of this guide, *containment* is used to cover both of these and is defined as the utilization of engineering controls either:

- to prevent the escape of materials hazardous to health into the surrounding workplace; or
- to prevent the contamination or degradation of compounds by the environment.

In some cases both the above objectives must be satisfied.

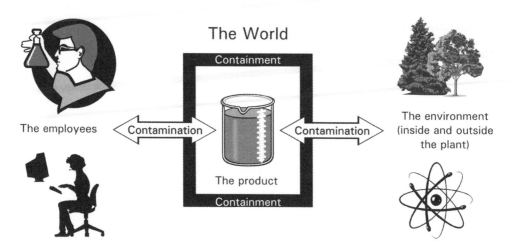

Figure 1.1 The purpose of containment

To date there has been a range of papers at conferences and in the literature which deal with specific issues on containment equipment and principles (For example, papers by Brooke (1998), Guest (1998), Maidment (1998), Russell *et al.* (1998) and Topping, Williams and Devine (1998) as detailed in Appendix 2.), but obviously their scope is limited. The purpose of this guide is therefore to embrace not only the many types of containment equipment available for use, but also the many issues that have to be considered by anyone selecting a particular containment system for a specific duty. These range from material properties and legal requirements through to validation and operational and maintenance needs.

Where is containment used?

The manufacturing industries to which the contents of the guide are relevant include those producing:

- speciality chemicals;
- chemical intermediates;
- agrochemicals;
- dyestuffs;
- pharmaceuticals, particularly those tasks involving:
  — active pharmaceutical ingredients;
  — pharmaceuticals in final dosage form.

Containment was originally developed for the nuclear industry and is also widely used in certain biological processing industries (e.g., production of vaccines) and in sterile process systems or those manufacturing sterile products. However, these are specialist areas, which require a specific approach or specialized design features. Thus, although they share many of the principles covered in this guide, they are outside its scope.

The hazards that may be presented by chemicals used in manufacturing processes or research laboratories, or by their resultant intermediates and products, include:

- flammability — the liability to burn if ignited;
- explosivity — the liability to explode if ignited;
- toxicity — the ability to damage human tissue or organs so as to cause illness or death;
- potency — the level of activity of a chemical.

The description **potent** is generally used of a chemical that produces significant physiological, toxicological or pharmacological effects on the body even if present in very small quantities. Such effects may be beneficial to a patient receiving a controlled dose of a drug but are liable to be harmful to workers manufacturing the drug who are exposed to uncontrolled quantities of it. Thus a substance that is of benefit to a patient could in effect become **toxic** to a worker. Since this guide is concerned primarily with protecting workers from the harmful effects of chemicals, for the sake of simplicity it uses the single term **toxicity** to cover potency as well.

Flammability is generally of lesser significance in determining the level of containment required. However, once a containment strategy has been selected according to the toxicity of the substances involved, as described in Chapters 6 and 7, additional precautions may be needed, as described in Chapter 8, if there is a risk of explosion, particularly if materials are handled that ignite on exposure to air, water or other substances.

The manufacture of chemicals and pharmaceuticals involves the transport of materials (solids, liquids or gases) into and out of process equipment. These activities have the potential to release dust or vapour into the workplace or to contaminate the product with unwanted materials present in the environment. Indeed, this guide could be described as a book about safe material transfer operations, in which releases are limited to within the allowable limits for that toxic material and in some cases to zero.

However, it must be stressed that adequate containment can be achieved only by taking an holistic approach to the problem, which will include not only equipment design and selection but also:

- objectives of the process (which must be clearly defined and understood);
- permitted personal exposure limits and other hygiene requirements;
- operational considerations;
- maintenance considerations;
- cleaning considerations;
- performance proving (**validation**).

What types of equipment are covered by this guide?

The design principles for containment equipment are not new. The early nuclear and biological industries had to develop closed systems for handling very hazardous materials. These included sophisticated glove-boxes and complex robot systems. The need to transfer hazardous materials into and out of such glove-boxes resulted in the development of the concept of the closed transfer port.

In the last 10–15 years, the process industries identified on page 2 have increasingly used a range of containment equipment, some of which is based on these principles but developed to meet transfer operations common to some or all of those industries.

This guide covers the range of containment equipment from simple air-flow control devices to enclosures that restrict exposures to well below a microgram per cubic metre averaged over a working day. For the most hazardous materials the need is identified for operators to be removed from the workplace and for robots to be used, for example, to transfer materials, but the detailed design of such specialized systems is outside the scope of the guide. As the manufacture of new toxic pharmaceutical products grows, so it becomes necessary to handle more compounds of increasing toxicity in the workplace. For this reason, and because the expectation of better employee protection and improved working procedures is growing, there will be an increasing demand for better containment systems and a better understanding of those systems. This guide covers a wide variety of such systems, suitable for materials of low to high toxicity, different scales of operation and different transfer steps.

The selection of a particular containment system for a particular transfer operation can be difficult because of the wide choice available. This guide provides a structured approach to the selection process.

Classification of toxic substances

Toxic compounds can be inhaled, absorbed through the skin or ingested. Therefore the approach to containment must consider all these routes of entry into the body.

Exposure Limits (OELs) (These are discussed further in Chapter 2) are set by a number of different bodies including legislative organizations, academic institutions and in-company groups. Their precise meaning and interpretation are normally published within the accompanying documentation and can differ depending on the group that assigns them. In general, exposure limits indicate the atmospheric concentration, averaged over a stated duration, to which a person may be exposed over a given period of time without adverse effect. They range in magnitude from several milligrams down to a few nanograms per cubic metre. In some cases exposures even smaller than these can add up over time to produce a harmful dose; this consideration will have been taken into account in setting the exposure limits. The containment methods described in this guide can be used to control exposures to below these levels.

The adverse effects that may occur as a result of exposure can vary and this guide uses the international classification system based on *Risk ('R') phrases*, summarized in Appendix 3, to describe them.

The relationship between exposure limits and the ability of different types of containment equipment to reduce exposures below them is discussed in Chapter 6 as part of the selection process for containment equipment.

Classification of containment systems

The terms 'primary' and 'secondary' containment have often been used to describe certain types of containment equipment. However, they have not been adopted in this guide because there are more than two fundamentally different methodologies for achieving containment.

The method recommended in the UK by the Health and Safety Executive (HSE) in their publication HSG193 (see Appendix 2 for full reference) is to identify the appropriate *control approach* based on a number of factors, including scale of operation and material properties as well as toxicity. This guide develops this approach by defining a number of *containment strategy* levels to meet increasing levels of containment and classifying the various equipment types accordingly.

Health and safety regulations in both the UK and the USA require engineering means to be used to control exposures rather than personal protective equipment. In considering how to design such systems, this guide distinguishes between *open handling*, where a substance is scooped or poured from one open container into another, and *closed handling*, where the transfer is effected through a closed system of pipes, ducts or directly connected vessels. In some cases, open handling takes place within a contained environment such as a glove-box.

Summary of remaining chapters

Chapters 2 and 3 provide general background information about the requirements that containment systems must meet. Chapter 2 outlines the legal requirements for containment as detailed in UK legislation, much of which is based on European Directives, and in the Codes of Federal Regulations issued by the Occupational Safety and Health Administration (OSHA) in the USA. Chapter 3 looks at the practical aspects of measuring and controlling exposures to hazardous substances.

Chapters 4 and 5 also provide background information but are more specifically concerned with typical material transfer operations in the process industries and the range of operational issues affecting the design and operation of containment systems.

Chapters 6 and 7 are concerned with selecting and describing the type of containment system required for a specific application. Chapter 6 defines five containment strategies and a rational methodology for selecting the one appropriate to the scale of use and the toxicity and other physical properties of the substances handled. Chapter 7 describes how the containment strategies are implemented in practice and includes details of the containment devices commonly available, some of which form quite complex systems.

Chapter 8 describes other important safety and operational aspects that must be considered after the appropriate control strategy has been identified. These include the safe disposal of waste materials and contaminated packaging and the safe treatment of process streams, such as exhaust gases and washing fluids, as well as various methods of preventing explosions if flammable materials are present.

To conclude the guide two important topics are covered. Chapter 9 describes the practicalities of implementing a containment system, including maintenance and operating procedures. Chapter 10 covers the performance validation and testing procedures to which most containment systems will be subject.

Appendices

In addition to the chapters described above, three Appendices are provided.

Appendix 1 provides a glossary of terms used in this guide. Terms introduced in this guide in **bold italics** are generally defined in this Glossary.

Appendix 2 lists the publications mentioned in the guide. In the case of HSE publications it provides the full references of publications referred to in the text by their HSE reference codes.

Appendix 3 lists the basic Risk (R) and Safety (S) Phrases specified in the HSE publication L100. These are used in the process, described in Chapter 6, for determining the containment strategy appropriate to a particular case.

Containment legislation

Purpose

The purpose of this chapter is to give an overview of current legislation in the UK and USA pertaining to the control and use of hazardous materials in the workplace. Parallel European legislation is referred to. A detailed explanation of terms is given in Chapter 3. Environmental legislation is considered in Chapter 8.

Contents

UK legislation

The **Health and Safety at Work etc. Act 1974** (HSW Act) is of overriding importance in all work situations in the UK. Not only does it impose duties in its own right, it also defines the authority of the Health and Safety Executive (HSE) to issue *Regulations*, which have the force of law. A few such Regulations are listed in Table 2.1.

The general duties and powers of the HSW Act provide protection to employees and to the public at large. The Act provides such wide protection that it not only requires employers to have regard to the health and safety of people at work, but also requires them to conduct their undertaking in . . .:

'such a way as to ensure, so far as is reasonably practicable, that persons not in [their] employment who may be affected thereby are not thereby exposed to risks to their health or safety'

There are four sets of Regulations relevant to the use of containment equipment. These are listed in Table 2.1 and described further in the sub-sections that follow.

Chemicals (Hazard Information and Packaging for Supply) Regulations 1994 (CHIP)

The **Chemicals (Hazard Information and Packaging for Supply) Regulations 1994** (CHIP) require suppliers to:

- identify the hazards of the chemicals they supply — this is called *classification*;
- inform those to whom they supply chemicals of the hazards of the chemicals supplied;
- package the chemicals safely.

These requirements are called *supply requirements* and are described further in Appendix 3.

The process steps to be followed when applying CHIP are shown in Figure 2.1, which is adapted from the HSE publication INDG181.

Table 2.1 Regulations relevant to containment

Regulation	Purpose
Chemicals (Hazard Information and Packaging for Supply) Regulations 1994 (CHIP) and subsequent amendments	Covers hazard identification, classification of substances, labelling and packaging
Control of Substances Hazardous to Health Regulations 1999 (COSHH)	Identifies the need and requirements for risk assessment, exposure control measures and other supporting activities
Supply of Machinery (Safety) Regulations 1992; Provision and Use of Work Equipment Regulations 1992	Together known as the *Machinery Directive* after the EU Directives 89/392/EEC and 98/37/EC, whose provisions they implement in the UK, they cover the marketing and certifying of machinery

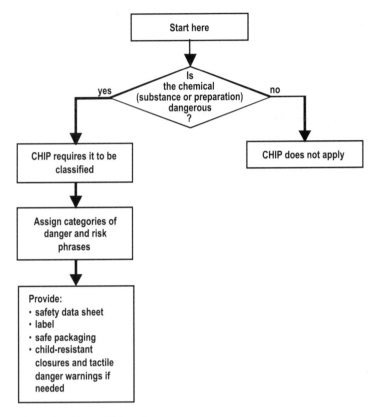

Figure 2.1 How CHIP works

Control of Substances Hazardous to Health Regulations 1999 (COSHH)

The **Control of Substances Hazardous to Health Regulations 1999** (COSHH) came into force on 25 March 1999. They revoke and replace:

- the original 1988 COSHH Regulations;
- the 1991, 1992, 1996, 1997 and 1998 COSHH Amendment Regulations;
- the 1994 COSHH Regulations;
- the Health and Safety (Dangerous Pathogens) Regulations 1981.

The Regulations and guidance on their implementation are described in an *Approved Code of Practice* (ACOP) published by the HSE as L5.

Substances hazardous to health

The definition of a *substance hazardous to health* is given in regulation 2 of COSHH. It covers virtually all substances, including preparations, capable of causing disease or other adverse health effects and applies to exposures to those substances arising from work

9

activities. A *substance* may be natural or artificial and may be present in solid, liquid, gaseous or vapour form. COSHH also applies to micro-organisms, although they are outside the scope of this guide.

COSHH identifies the following five categories:

(a) substances in Part 1 of the Approved Supply List (within the meaning of the CHIP Regulations, currently L115) in the *very toxic, toxic, harmful, corrosive* or *irritant categories*;

(b) substances for which the Health and Safety Commission (HSC) has approved a *maximum exposure limit* or an *occupational exposure standard* (described later in this chapter);

(c) biological agents, i.e., micro-organisms, cell cultures or human endoparasites (including genetically modified micro-organisms) capable of causing any infection, allergy, toxicity or other human health hazard;

(d) any dust at a substantial concentration in air (i.e., greater than or equal to $10 \, \text{mg m}^{-3}$ for total *inhalable dust* or $4 \, \text{mg m}^{-3}$ for *respirable dust*, in both cases measured as 8-hour *Time-weighted Averages* (TWA) values);

(e) any substance not in (a) to (d) above but which creates a health hazard comparable to any of them.

Specifically excepted from COSHH are exposures to lead and asbestos to the extent that these substances are covered respectively by the **Control of Lead at Work Regulations 1998** and the **Control of Asbestos at Work Regulations 1987**, as amended.

Also outside the provisions of COSHH are substances:

• hazardous by virtue of radioactive, explosive or flammable properties;
• hazardous solely because of high or low temperature or high pressure;
• administered as part of medical treatment;
• used below ground in mines to which the **Mines and Quarries Act 1954** applies.

Assessment of Health Risks (regulation 6)

This regulation prohibits employers from carrying out any work liable to expose employees to substances hazardous to their health, unless a suitable and sufficient assessment has been made of:

• the risks created by that work;
• the steps needed to comply with COSHH in respect of it.

Assessments should be reviewed regularly and in any case if:

• their validity becomes suspect; or
• significant changes take place in the work or the way it is carried out.

Assessments should be sufficient to enable employers to make informed and valid judgements about health risks. They must be undertaken by a competent person and should consider what substances employees are likely to be exposed to. They should also:

- consider the potential effects of such substances, taking into account:
  — their quantity and form;
  — the amount to which employees may be exposed;
- compare such exposure to any relevant published standards;

To derive the data that will form the basis of an adequate assessment, the competent assessor will consider:

- past experience;
- previous records, where they exist;
- information on toxicity gleaned from statutory labelling requirements and manufacturers' safety data sheets (which must be supplied as a requirement of CHIP);
- findings of surveys and sampling in the workplace;
- Toxicity Reviews published by the HSE (in their 'TR' series);
- journals;
- any other sources of relevant information produced by trade associations, etc.

Except in the very simplest of situations, it is necessary for assessments to be in written form. This will enable employers to demonstrate readily that:

- all factors have been considered;
- they have adequate knowledge on which to base the control measures required to achieve compliance with COSHH;
- they can show continuity of effort and achievement in dealing with health risks associated with the work.

Prevention or Control of Exposure to Substances Hazardous to Health (regulation 7)

The first duty of every employer is to *prevent* the exposure of employees to substances hazardous to their health. Only where prevention is not reasonably practicable may the employer consider *controlling* exposure as a legitimate second option.

Adequate control must be secured in the first instance by measures other than *Personal Protective Equipment* (PPE), e.g., by enclosure of the process or by the use of *Local Exhaust Ventilation* (LEV). Where such control measures are found to be inadequate by themselves, then, and only then, is the employer able to resort legitimately to the provision of suitable PPE. Any item of PPE provided by the employer must meet the requirements of any relevant EU Product/Design Directives as listed in Schedule 1 of the **Personal Protective Equipment at Work Regulations 1992**. In each case it is the employer who has to establish what is 'reasonably practicable'.

Certain substances have been given *Maximum Exposure Limits* (MELs). Where such limits exist, control of exposure, so far as inhalation of those substances is concerned, can be treated as adequate only if the level of exposure is reduced so far as is reasonably practicable and in any case *below* the listed MEL.

Certain other substances have been given approved *Occupational Exposure Standards* (OESs). Again, so far as inhalation of these substances is concerned, control is treated as adequate if the occupational standard is not exceeded or, where it is exceeded, the employer

11

identifies why it is and takes appropriate action 'as soon as reasonably practicable' to remedy the situation.

MELs and OESs are collectively known as *Occupational Exposure Limits* (OELs). A complete list of OELs can be found in the HSE's annually revised publication EH40, which also contains information on setting OELs, applying OELs, new limits, reviews and revisions, mixed exposures, skin absorption and respiratory sensitizers, as well as a technical supplement and a number of appendices.

For many of the substances listed in EH40, there are two exposure limits set, which reflect variations in the effects that may be expected, depending on the nature of the substance and the pattern of exposure.

Some effects will occur following prolonged or accumulated exposure. The *long-term* (8-hour TWA) exposure limit is intended to be used to control such effects by restricting the total intake by inhalation over one or more work shifts. Other effects may be apparent following shorter exposures. *Short-term* exposure limits (usually averaged over 15 minutes) may be applied to control these effects. Where long-term limits also apply, the short-term limits restrict the magnitude of excursions above the average concentration that may be permitted during longer exposures.

OESs and MELs are set on the recommendations of the HSC's Advisory Committee on Toxic Substances and its Working Group on the Assessment of Toxic Chemicals. Following detailed review and consideration of all the relevant information, these committees consider, in accordance with specific rules:

- what type of limit should be set;
- the level at which the limit should be set.

Any *Respiratory Protective Equipment* (RPE) that needs to be provided has to be suitable for its intended purpose and should be of a type that conforms to any UK legislation that implements the relevant EU Product Directives concerning design and manufacture require-ments, e.g., the **Personal Protective Equipment (EC Directive) Regulations 1992**. If there is no such applicable legislation, the RPE should be approved by or conform to a standard approved by the HSE.

Carcinogens

Carcinogens are substances liable to cause cancer if ingested, inhaled or otherwise taken into the body. The following hierarchy of controls should be applied where it is not reasonably practicable to prevent exposure to carcinogens by the use of alternative substances or processes:

(1) totally enclose the process or system;
(2) minimize, suppress and contain the generation, spillage or leakage of carcinogenic dusts, fumes and vapours through the use of appropriate plant, processes and work systems;
(3) minimize the quantities of carcinogens on site and the number of persons likely to be exposed;
(4) prohibit eating, drinking and smoking in contaminated areas;
(5) provide facilities for personal washing and cleaning of premises;

(6) designate contaminated areas and installations (including suitable marking);

(7) safely store, handle and dispose of carcinogens (using closed and clearly labelled containers among other precautions).

If the failure of a control measure results in the escape of a carcinogen into the workplace, the employer must ensure that:

- only those personnel necessary for the repairs, etc., are authorized to enter the affected area;
- those personnel are provided with suitable PPE;
- employees and anyone else who may be affected by the escape are informed immediately.

Use, Maintenance, Examination and Test of Control Measures (regulations 8 and 9)

Employers providing control measures, including PPE, must take 'all reasonable steps' to ensure that they are properly used or applied.

Employees must 'make full and proper use of control measures', including PPE, that are provided to comply with COSHH requirements. If employees discover any defect in any control measure provided, they must report it at once to their employer. How to recognize defects and understand their implications and significance is an important part of the training regime, which has to be provided by the employer under regulation 12.

The employer has to ensure that all the control measures provided are maintained in an efficient state, in efficient working order and in good repair. Where possible, they should usually be checked at least once every week. Any PPE used must additionally be kept clean and stored in suitable accommodation, which the employer must provide.

LEV must be thoroughly examined and tested at least once in every 14 months (in certain specified cases the interval between consecutive examinations is shorter, as detailed in Schedule 4 of the COSHH Regulations (Schedule numbers here and elsewhere in this chapter refer to the COSHH Regulations 1999. The schedules are numbered differently in previous editions of the Regulations)). Other engineering controls must be thoroughly examined and tested at suitable intervals.

Where RPE is provided, employers must ensure that it is examined at suitable intervals and, where appropriate, tested. Thorough examination and testing should be carried out at least every three months, though in many cases more frequent testing will be needed.

Suitable records of examinations and tests, and of the repairs found to be necessary and carried out thereafter, must be kept available for inspection for at least five years.

Monitoring Exposure at the Workplace (regulation 10)

Where employees are exposed to hazardous substances and the assessment under regulation 6 shows that monitoring is required, that exposure must be monitored by a suitable procedure. For some specific substances or processes, Schedule 5 of the COSHH Regulations sets out the minimum frequency at which monitoring must be carried out. Records of monitoring must be kept for at least five years. Where the record of monitoring is representative of the personal exposure of identifiable employees, the period for maintaining the records is extended to 40 years.

Health Surveillance (regulation 11)

Where it is appropriate to protect the health of employees exposed to substances hazardous to their health, the employer must ensure that such employees are subject to suitable health surveillance including, where necessary, biological monitoring. For the substances and processes specified in Schedule 6 of the COSHH Regulations, health surveillance is appropriate unless the exposure is not significant.

Surveillance may also be appropriate in other cases where exposure to hazardous substances is such that:

- an identifiable disease or adverse health effect may occur under the particular conditions of work prevailing;
- valid techniques exist to detect such conditions and effects.

The employer must keep records of surveillance in respect of each employee for at least 40 years after the last date of entry. This requirement still applies where companies cease to trade, in which case the records must be offered to the HSE.

Information, Instruction and Training for Employees (regulation 12)

Employees exposed to substances hazardous to their health must be provided with such information, instruction and training as is suitable and sufficient for them to know the health risks associated with their exposure to the substances with which they work. They should also be familiar with the precautions necessary for their protection.

The information provided to employees must include the results of monitoring and the collective, but not individual, results of any health surveillance undertaken.

Where monitoring results indicate that MELs have been exceeded, employees, or their representatives, should be told at once.

All persons carrying out any work in connection with COSHH (assessment, monitoring, health surveillance, training, examining control equipment, etc.) must have had the necessary information, instruction and training. This applies whether or not the persons concerned are employees of the employer for whom they are undertaking this work.

Miscellaneous provisions

Apart from the main provisions governing the protection of employees' health, as outlined above, COSHH deals with a number of other related matters, as follows:

- where duties are placed on employers to protect their employees, they are, so far as is reasonably practicable, under a similar duty in respect of persons they do not employ (whether they are at work or not) who may be affected by their handling of hazardous substances at the workplace. However, there is no duty on employers to provide health surveillance for non-employees and the provisions relating to monitoring, exposure, information and training apply to non-employees only where they are on the premises where the work is carried out.
- self-employed persons are covered by the Regulations as if they are both employers and employees, although they do not have to monitor exposure or carry out health surveillance;

- the use of certain substances (listed in Schedule 2 of the COSHH Regulations) is prohibited for certain purposes, e.g., sand or other substances containing free silica may not be used as an abrasive for blasting articles;
- any provisions required under environmental protection or public health legislation must not be prejudiced by any requirement under COSHH;
- the defence of *due diligence* applies to a contravention of these regulations, i.e., it is a defence for any person to prove that (s)he had taken all reasonable precautions, and thereby exercised all due diligence, to avoid committing the offence.

Control of substances not assigned an occupational exposure limit

EH40 lists OELs for over 500 substances. Absence of a substance from EH40 does not indicate that it is safe. Indeed, many compounds being investigated as efficacious pharmaceuticals are highly toxic in their pure form. In-house limits, therefore, need to be set for intermediates and it is common practice in the pharmaceutical industry for late-stage intermediates of potent compounds to be assigned the same OEL as the final product unless specific information is available.

Some bodies, such as the Association of the British Pharmaceutical Industry, have produced guidance (e.g., ABPI, 2000) on setting in-house limits for pharmaceuticals. The Chemical Industries Association have produced guidance (CIA, 1997) on allocating occupational exposure bands and more specific guidance (CIA, 1993) for dyestuff users on hazard classification and the safe handling of colourants.

In Chapter 6 of this guide, a method of establishing the control strategy for compounds without an OEL is given, which is in line with the recommendations of the HSE as described in HSG193.

Machinery directive

There are two groups of law derived from the **EU Machinery Directive**:

- One deals with what manufacturers and suppliers of *new* machinery have to do. This can be called the *supply law*. The UK legislation most widely used in this respect is the **Supply of Machinery (Safety) Regulations 1992**, which require manufacturers and suppliers to ensure that machinery:
  — is safe when supplied; and
  — carries the CE mark.
- The other deals with what the users of machinery and other equipment have to do. This can be called the *user law*. The piece of legislation that applies most widely here is the **Provision and Use of Work Equipment Regulations 1992**. These require employers to:
  — provide the right kind of safe equipment for use at work;
  — ensure that it can be used correctly;
  — keep it maintained in a safe condition.

Anyone who buys new equipment (including machinery) is also required under this law to check that the equipment complies with all the other relevant supply law.

The user law includes other requirements, but this guide does not deal with them.

What is meant by 'machinery'?

A *machine* is normally regarded as being a piece of equipment that has moving parts and, usually, some kind of drive unit. Many types of containment equipment are classified as *machinery*.

What does the manufacturer have to do?

Manufacturers must make sure that the machines they make are safe. They will do this by:

- finding out about the health and safety hazards (trapping, noise, crushing, electrical shock, dust, vibration, etc.) that are likely to be present when the machine is used;
- assessing the likely risks;
- designing out the hazards that result in risks; or, if that is not possible:
- providing safeguards, e.g.:
  — guarding dangerous parts of the machine;
  — providing noise enclosures for noisy parts; or, if neither of these is possible:
- using warning signs on the machine to warn of hazards that cannot be designed out or safeguarded (e.g., 'NOISY MACHINE — WEAR HEARING PROTECTION');

  Manufacturers must also:

- keep information, explaining what they have done and why, in a technical file;
- fix a CE marking to the machine where appropriate, to show that they have complied with all the relevant supply laws;
- issue a *declaration of conformity* for the machine;
- provide the buyer with instructions to explain how to install, use and maintain the machinery safely.

  It is important to realize that the above definition of 'safe' is based on the prevention of injuries caused by moving parts, etc. It is not concerned with the hazards of toxic materials. The purchaser's responsibility under these Regulations is to specify the requirements for the performance of the machinery in *mechanical* terms.

Before buying a new machine

Anyone buying a machine should first think carefully about:

- where and how it will be used;
- what it will be used for;
- who will use it (skilled employees, trainees);
- what risks to health and safety might result;
- how well it controls health and safety risks compared with machines supplied by other manufacturers.

  Such considerations can help the purchaser to decide which machine may be suitable, particularly when buying a standard machine 'off the shelf'.

  Those buying a more complex or custom-built machine should discuss their requirements with potential suppliers. Suppliers can often provide useful advice on the options available.

Anyone ordering a custom-built machine should use the opportunity to work with the supplier to design out any potential causes of injury and ill health. Time spent at the design stage on agreeing the safeguards needed to control health and safety risks could save the purchaser time and money later.

Note: Sometimes machinery is supplied via another organization, e.g., an importer, rather than direct from the manufacturer. In such cases, this organization would be referred to as the 'supplier'.

When placing the order, the purchaser should *specify* in writing that the machine should be safe in accordance with recognized codes.

Before accepting the machine after purchase, purchasers should:

- look for the CE marking;
- check that they have a copy of the declaration of conformity and a set of instructions that they and their employees can understand on how the machine should be used.

Most important of all, they should *satisfy themselves that it is safe*. Guidance on how to do so is given in the HSE publication INDG271.

Technical files

Manufacturers are required to draw up technical files for the machinery they make. These files must include:

- drawings of the machinery and its control circuits;
- the specifications and standards used in the design;
- information about relevant test results.

Technical reports and certificates from other organizations may also be included, as may any other relevant information.

Technical files demonstrate how machinery meets relevant essential health and safety requirements and, as such, are useful for manufacturers and for the national enforcing authorities. However, manufacturers are not obliged to make the contents of technical files available to other suppliers or to the eventual users of the machinery.

US legislation

The Occupational Safety and Health Administration (OSHA) within the US Department of Labor issues Codes of Federal Regulations, of which Title 29 (Labor) includes Part 1910 entitled **Occupational Safety and Health Standards**. This part, authorized by the **Occupational Safety and Health Act** of 1970 (84 Stat. 1593), promulgates standards...

'reasonably necessary or appropriate to provide a safe or healthful employment and place of employment.'

The standards contained in this part apply with respect to employment performed in a state, District of Columbia, territories and outer continental shelf lands. They do not apply to many Federal agencies.

Except for certain appendices designated as advisory, all the provisions of part 1910 need to be observed, including sections such as 29 CFR 1910.95, **Occupational Noise Exposure** or 29 CFR 1910.133, **Eye and Face Protection**. However, the sections most relevant to containment devices are listed in Table 2.2 and described further in the remaining sections of this chapter.

29 CFR 1910.134 – Respiratory Protection

This section states:

'*In the control of those occupational diseases caused by breathing air contaminated with harmful dusts, fogs, fumes, mists, gases, smokes, sprays or vapors, the primary objective shall be to prevent atmospheric contamination. This shall be accomplished as far as feasible by accepted engineering control measures (for example, enclosure or confinement of the operation, general and local ventilation, and substitution of less toxic materials). When effective engineering controls are not feasible, or while they are being instituted, appropriate respirators shall be used pursuant to this section.*'

To comply with this section the employer must develop and implement a site-specific written respiratory protection programme that includes:

- procedures for selecting respirators;
- medical evaluations;
- fit testing;
- procedures for use;
- procedures and schedules for cleaning, disinfecting, storing, inspecting, repairing, discarding and otherwise maintaining respirators;

Table 2.2 Sections of CFR Title 29 Part 1910 relevant to containment

Section	General purpose
29 CFR 1910.134 Respiratory Protection	Governs use of respirators when engineering controls are not adequate to control occupational diseases caused by airborne contaminants.
29 CFR 1910.1000 Air Contaminants	Gives exposure limits for air contaminants, mineral dusts and nuisance dusts.
29 CFR 1910.1200 Hazard Communication with appendices	The purpose of this section is to ensure that the hazards of all chemicals produced or imported are evaluated and that the information is transmitted to all those who need it.
29 CFR 1910.1450 with appendices	Governs the use of hazardous chemicals in laboratories.

- training in proper use and the hazards of misuse;
- procedures for evaluating the effectiveness of the programme;
- provisions for updating the programme as necessary to reflect changes in the workplace.

29 CFR 1910.1000 – Air Contaminants

This section includes three tables, Z-1, Z-2 and Z-3, which contain *Permissible Exposure Levels* (PELs) for various listed chemicals. PELs define the maximum airborne concentrations of those chemicals to which personnel may be exposed and take the form of one or more of:

- 8-hour *Time-weighted Average* (TWA) values, maximum permissible exposures averaged over an 8-hour working day;
- *ceiling levels*, which may not be exceeded at any time;
- *acceptable ceiling concentrations*, which may not be exceeded except for designated time periods and then only if the TWA concentration over those periods remains below a designated *acceptable maximum peak* level.

The three tables list PELs as follows:

- Table Z-1 provides 8-hour TWA values and some ceiling values for airborne materials in parts per million (ppm) and mg m^{-3};
- Table Z-2 provides 8-hour TWA values, acceptable ceiling levels and acceptable maximum peak levels for toxic and hazardous substances in ppm and mg m^{-3};
- Table Z-3 provides 8-hour TWA values for mineral dusts, including inert and nuisance dusts, in millions of particles per cubic foot (mppcf) and mg m^{-3}.

29 CFR 1910.1000 also includes methods for calculating whether exposures to specific compounds or mixtures of components are acceptable.

To achieve compliance...

'administrative engineering controls must first be determined and implemented whenever feasible. When such controls are not feasible to achieve full compliance, protective equipment or any other protective measures shall be used to keep the exposure of employees to air contaminants within the limits prescribed in this section. Any equipment and/or technical measures used for this purpose must be approved for each particular use by a competent industrial hygienist or other technically qualified person. Whenever respirators are used, their use shall comply with 1910.134.'

29 CFR 1910.1200 – Hazard Communication

Paragraph (a)(1) of this section reads:

'The purpose of this section is to ensure that the hazards of all chemicals produced or imported are evaluated, and that information concerning their hazards is transmitted to employers and employees.'

Paragraph (a)(2) states:

'This occupational safety and health standard is intended to address comprehensively the issue of evaluating the potential hazards of chemicals, and communicating information concerning hazards and appropriate protective measures to employees, and to preempt any legal requirements of a state, or political subdivision of a state, pertaining to this subject. Evaluating the potential hazards of chemicals, and communicating information concerning hazards and appropriate protective measures to employees, may include, for example, but is not limited to, provisions for: developing and maintaining a written hazard communication program for the workplace, including lists of hazardous chemicals present; labelling of containers of chemicals in the workplace, as well as of containers of chemicals being shipped to other workplaces; preparation and distribution of material safety data sheets to employees and downstream employers; and development and implementation of employee training programs regarding hazards of chemicals and protective measures.'

This section applies to chemicals that are known to be in the workplace and to which employees could be exposed under normal conditions or in a foreseeable emergency. It applies to a limited extent to laboratories and warehouses (where employees handle chemicals only in sealed containers).

The section includes the following definitions of terms used in the extract above:

"Hazardous chemical' means any chemical which is a physical hazard or a health hazard.'

"Health hazard' means a chemical for which there is statistically significant evidence based on at least one study conducted in accordance with established scientific principles that acute or chronic health effects may occur in exposed employees. The term 'health hazard' includes chemicals which are carcinogens, toxic or highly toxic agents, reproductive toxins, irritants, corrosives, sensitizers, hepatotoxins, nephrotoxins, neurotoxins, agents which act on the hematopoietic system, and agents which damage the lungs, skin, eyes or mucous membranes. Appendix A provides further definitions and explanations of the scope of health hazards covered by this section, and Appendix B describes the criteria to be used to determine whether or not a chemical is to be considered hazardous for purposes of this standard.'

"Physical hazard' means a chemical for which there is scientifically valid evidence that it is a combustible liquid, a compressed gas, explosive, flammable, an organic peroxide, an oxidizer, pyrophoric, unstable (reactive) or water-reactive.'

Manufacturers and importers are responsible for evaluating chemicals to determine whether they are hazardous. Compounds that are deemed to be hazardous must:

- be labelled appropriately;
- be provided with a **Material Safety Data Sheet** (MSDS) complying with the requirements of this section.

Employees must be provided with effective information and training on hazardous chemicals in their work areas. The information required is specified in paragraph (h) of 29 CFR 1910.1200 and includes:

- details of operations and procedures in their work areas where hazardous chemicals are present;
- the location of written hazard information.

Training required includes:

- methods and observations that may detect the presence or release of a hazardous chemical (e.g., monitoring);
- physical and health hazards of the chemicals;
- measures to be taken to protect employees from the hazards, including procedures, work practices and PPE.

Section 1200 has four appendices (A, B, D and E — Appendix C was withdrawn in 1996), which are important in implementing its requirements:

- Appendix A defines the following terms as applied within the section:
  — *carcinogen*;
  — *corrosive*;
  — *highly toxic*;
  — *irritant*;
  — *sensitizer*;
  — *toxic*.
- Appendix A also includes a list of target organ effects, a categorization of effects produced by various chemicals on specified organs in the human body;
- Appendix B describes general evaluation criteria to be used to determine hazards, which include:
  — carcinogenicity;
  — human data;
  — animal data;
  — adequacy and reporting of data;
- Appendix B also points out that:

'Hazard evaluation is a process which relies heavily on the professional judgment of the evaluator, particularly in the area of chronic hazards. The performance-orientation of the hazard determination does not diminish the duty of the chemical manufacturer, importer or employer to conduct a thorough evaluation, examining all relevant data and producing a scientifically defensible evaluation.'

- Appendix D defines the terms *trade secrets* and *secrecy*;
- Appendix E gives guidance to employers on how to achieve compliance with the requirements of this section.

29 CFR 1910.1450 – Occupational Exposure to Hazardous Chemicals in Laboratories

This section applies to all employees who use hazardous chemicals in laboratories. It requires that the exposure of laboratory workers to OSHA-regulated substances shall not exceed PELs.

The section also requires that a *chemical hygiene plan* be instituted where hazardous chemicals are used. Such a plan should include:

- standard operating procedures;
- criteria used to determine control measures;
- a requirement to ensure that fume hoods and protective equipment are functioning properly;
- provisions for information and training;
- circumstances under which prior approval is needed for specified operations;
- provisions for medical consultation and examination;
- responsibilities for implementing the plan;
- provisions for special protection for *select carcinogens* (as defined in the section), reproductive toxins and substances that have a high degree of acute toxicity, including, where appropriate:
  — segregation;
  — use of containment devices;
  — safe removal of wastes;
  — decontamination procedures.

Appendix A of 29 CFR 1910.1450 is non-mandatory but may be used as a guide in the development and application of a chemical hygiene plan.

Conclusion

This chapter has outlined the legislation in the UK and USA governing the identification and control of risks that may arise in processes using hazardous chemicals and associated equipment. The next chapter considers how compliance with these requirements may be achieved in practice, by describing how an occupational hygienist will carry out such a programme of identification and assessment and what forms of control are available.

Occupational hygiene aspects of containment

3

Purpose

The purposes of this chapter are twofold:

- to define commonly used terms relating to hazardous materials;
- to outline the occupational hygienist's approach to the identification and control of hazards.

Contents

Introduction

An engineer designing a process plant will need to identify whatever means are appropriate to prevent or control exposure of people to hazardous materials. Experience has shown that, in the past, appropriate methods of control were not well understood and many industries have therefore had to resort to retrofitting, which is both difficult and expensive. Traditionally, occupational hygienists have had little involvement in plant design or modification. However, this is now changing and design engineers recognize the value of hygienists' experience and the help they can offer.

Occupational hygiene is concerned with the anticipation, recognition, evaluation and control of those environmental factors (e.g., chemical, physical and microbiological agents and ergonomics) that may affect the health and well-being of individuals at work or in the community. This guide is focusing on chemicals, but other hazards such as noise and ergonomics can be equally important factors in design.

With the increasing rapidity of scientific progress in medical and biological fields, research-based pharmaceutical companies are able to develop and market increasingly more potent and selective products. These products require only very small doses to produce an effect and consequently can have adverse effects on employees exposed even to very low concentrations. With the advent of these compounds, the experience gained by occupational hygienists in existing facilities can contribute to the anticipation of potential risks in new plants. Chapters 6 and 7 have incorporated some of this experience, but there is still considerable work that needs to be undertaken to classify all forms of control.

Assessment

An assessment is required to understand the hazards associated with chemicals and the risks that they pose. It is important to understand the distinction between the two:

- the *hazard* is the potential harm that can arise from use of a chemical; it reflects the inherently dangerous properties of the chemical, which are fixed;
- the *risk* is the likelihood that the hazardous properties of a chemical will cause actual harm to people or the environment; it is dependent on the circumstances of use.

For example, sodium cyanide is a very hazardous material but if it is totally contained then the risk is low. Conversely, a low-hazard material such as starch may pose a significant risk if large quantities are handled with no controls.

One of the primary activities of an occupational hygienist is the assessment of risks associated with workplace activities. The Health and Safety Executive (HSE) publish a wide range of guidance, of which HSG97 provides a useful 'step-by-step-guide' to COSHH assessments. A *suitable and sufficient assessment* comprises a detailed review of procedures and practices associated with the handling of hazardous substances, with the object of ensuring that unnecessary exposures are eliminated or controlled. Regulation 6, as outlined in Chapter 2, has covered the basic requirements of an assessment and the following provides more details on how to carry one out.

An assessment of the risks to health posed by chemicals requires:

- collection of information on the:
  — tasks;
  — substances;
  — existing controls;
- observation of the activities;
- discussions with personnel carrying out the work;

The assessment provides a systematic approach for managing the risks from chemicals and identifies priorities for:

- controls (see page 10);
- maintenance, examination and testing of existing controls;
- air monitoring.

It also identifies the needs for health surveillance and training.

Work activities

Chemical risk assessments are normally carried out on specific work activities. It is necessary to know the number of people involved in the operations and the total number of people that may undertake the task. In addition, it is necessary to establish the frequency and duration of the task to be assessed. All tasks should be considered, including cleaning and maintenance as well as routine activities. Deviations from normal practice should also be considered if they have the potential to increase the likelihood of loss of containment.

Substances

It is necessary to find out what substances are present or generated in the workplace and where they are used, processed, handled or stored. This is usually done by checking inventories and considering the processes to establish any intermediates, excipients, by-products or finished products and any waste products that may be generated. All activities, whether associated with production processes, laboratories, warehouses, offices or any other location, should be included in the investigation. The quantities handled of each hazardous substance identified should be recorded along with any occupational exposure limits (see page 26).

Tests should be made on the substances identified in order to establish other toxicological effects, and the organs where they occur, such as:

- acute toxicity;
- chronic toxicity;
- carcinogenicity;
- mutagenicity (ability to change genes).

These tests and human exposure data are used to provide information for *Material Safety Data Sheets* (MSDSs), which can be obtained from suppliers and, in some cases, from in-house sources. It is essential to establish that an MSDS is up-to-date before making use of information it contains.

Within the EU, there is a standard 16-section format for MSDSs. Requirements in the USA are specified in 29 CFR 1910.1200(g). The requirements in either case include:

- identification;
- composition;
- toxicological and hazard identification;
- handling;
- storage;
- stability;
- exposure limits (described in more detail in the next section) and controls;
- labelling requirements;
- fire safety information.

The labelling information required in the EU will include pictograms that show the type of hazard (e.g., toxic, irritant, flammable). It will also give more specific information in the form of:

- Risk (R) phrases, which are standard phrases giving simple information about the hazards of a chemical in normal use;
- Safety (S) phrases, which provide advice on the safety precautions that may be appropriate when using the chemical.

The basic R-phrases and S-phrases approved for use in the UK are defined in L100 and listed in Appendix 3 of this guide.

Exposure limits

As mentioned in Chapter 2, the HSE in the UK publish a new edition every year of EH40, which contains a list of *Occupational Exposure Limits* (OELs), comprising *Maximum Exposure Limits* (MELs) and *Occupational Exposure Standards* (OESs), for use in assessing and ensuring compliance with the COSHH Regulations. In the USA, *Permitted Exposure Levels* (PELs) are defined in Tables Z-1 to Z-3 of 29 CFR 1900.1000.

The values quoted in EH40 and 29 CFR 1910.1000 for exposure limits relate only to exposure via inhalation. Consequently, when assessments are undertaken, other routes of exposure, e.g., skin contact and swallowing, also have to be considered.

Controls

A variety of controls (see page 32) may already be in place to prevent or limit exposure and it will be necessary to evaluate these as part of the assessment.

It is necessary to determine whether the measures are:

- effective, e.g.:
  — does the exhaust ventilation actually capture the contaminant?
  — do the gloves provide an effective barrier?
- reliable, e.g.:
  — might the ventilation break down?
  — might the trip switch fail?

— might the gloves tear?
— properly maintained, e.g.:
- does the ventilation pull as much air as it should?
  — are the flexible connectors in good condition?
  — are the connections leak-tight?
  — are the ducts blocked?
  — are the worn gloves replaced?
  — are the gauges regularly calibrated?

Loss of control, even for a short period of time, as a consequence of incorrect use, poor maintenance, etc., can have a dramatic effect on the degree of protection obtained.

For a general control factor F, the dose received from an airborne concentration x when the control measure is not working (e.g., the respirator protection is not worn or not effective) for $p\%$ of the time is given by:

$$D_p = \left(1 + \frac{(E-1)p}{100}\right)\frac{x}{F}$$

For example, if a respirator with a protection factor of 1000 (i.e., one that reduces the concentration inhaled by the operator to 0.001 of the atmospheric concentration in the workplace) is worn throughout the entire shift (i.e., $p = 0$), the dose received is given by:

$$D_0 = \left(1 + \frac{999 \times 0}{100}\right)\frac{x}{1000} = \frac{x}{1000}$$

However, if the respirator is not worn for 2% of the time (i.e., $p = 2$), then the dose received is given by:

$$D_{2\%} = \left(1 + \frac{999 \times 2}{100}\right)\frac{x}{1000} = 20.98\frac{x}{1000}$$

Thus the failure to wear the respirator for just 2% of the time has produced an increase of 20-fold in the dose received and reduced the respirator's effective protection factor from 1000 to a little under 48.

Observations and discussions with personnel

Some of the information that must be established when assessing the need for exposure controls can be obtained through direct observations and discussions with site personnel. The information needed, and the sorts of questions that may identify it or factors that should be borne in mind, are as follows:

- the prevailing operating conditions:
  — are all processes running?
  — is overtime being worked?
- who may be exposed:
  — is exposure restricted to those people carrying out the work or could it also affect supervisors, visitors, maintenance workers, etc.?
  — are any of the people exposed particularly vulnerable, e.g., adolescents, pregnant women or employees with pre-existing medical conditions?

- the pattern of exposure:
  — is it continuous or intermittent?
  — is it frequent or occasional?
  — does it occur at certain times, e.g., when loading a batch of raw material?
  — is it confined to short periods?
  — can its duration exceed the normal eight-hour shift length?
- the working practices — small differences in working practices can make huge differences to exposure, e.g.:
  — a person's working position relative to the source of exposure affects the amount of contaminant breathed in;
  — the procedures used for handling powders have a major effect on the amount of dust generated;
  — incorrect positioning of Local Exhaust Ventilation (LEV) or working outside its operating range can render it ineffective;
  — the way in which gloves are donned and removed can cause contamination of the inside surface, leading to absorption of contaminants through the skin;
  — washing and changing at breaks can minimize ingestion of contamination on the skin or clothing;
- any incidents or concerns about ill health:
  — incidents, even those that did not cause any actual harm, may suggest other incidents that could occur;
  — some employees may have allergies or be especially susceptible to harm in some other way.

An assessment of exposure must take into account the reality of working practices. If employees take short cuts for convenience, make 'improvements' without fully appreciating the consequences or reject protective clothing that is uncomfortable, these facts must be recorded and used in the estimation of the risk.

It is necessary to identify operations or situations that create the opportunity for exposure. Eating and drinking in the workplace, poor personal hygiene and the application of cosmetics in the workplace may lead to unnecessary exposure to chemicals. Some chemicals, identified in EH40 by the code 'Sk' or by a 'skin' designation in Table Z-1 of 29 CFR 1910.1000, can be absorbed through the skin. Some of these produce dermatitis through only brief contact with the skin. Others (e.g., glycol ethers and phenol) pass through the skin and cause damage to various organs of the body; indeed, this may be the primary route of entry under certain circumstances.

Dermal exposure estimation will depend on the surface area of contact, quantity on the skin and degree of absorption. While gases can pass through the skin to some degree, dermal exposure results mostly from contact with solids and liquids. The potential for skin contact depends on:

- level of activity;
- likelihood of contact;
- frequency of contact;
- surface area in contact;

- physical properties of the material that determine retention;
- use of protective equipment;
- frequency of washing.

Exposure by inhalation can occur where dusts, fumes, gases or vapours are released into the workplace air, e.g.:

- in production:
  — dusts from handling of powders or dusty materials during dispensing, weighing, milling, blending, loading into drums, etc.;
  — vapours from evaporation of liquids, e.g., solvents and gases formed by process reactions or from leaking supply lines or valves;
- in maintenance:
  — dusts from grinding, sawing or drilling solid materials;
  — fumes from molten metals during welding;
  — dusts or vapours from servicing containment devices;
- in laboratories:
  — vapours from the use of solvents or other volatile compounds;
  — chemicals in the form of powders.

By identifying situations such as these, the source of exposure can be tracked down even if the contaminant itself is invisible. If there are no sources, then by definition there can be no risk. If there are sources it is necessary to assess whether exposure is likely to occur in practice or even whether it is already occurring.

During the observation process, evidence of exposure should be sought. There are a number of signs that will indicate the probable extent of exposure:

- Deposits of settled dust on horizontal surfaces indicate that the dust has been airborne at some time. The depth of deposit depends on both the concentration of material in the air and the time since the surface was last cleaned. At a concentration of $1\ \mathrm{mg\,m^{-3}}$ a light film of dust would be expected to appear on any surfaces that are not cleaned frequently. Such situations often provoke complaints from employees about nuisance from dusts, e.g., contamination of clothing, hands and hair. If these occur, then there is also a likelihood of ingestion.
- Many organic solvents can be detected by smell, often at concentrations less than their exposure limits. Odour thresholds are available from several information sources, which should be referenced when required. However, odour is not a reliable indicator as some chemicals (e.g., hydrogen sulphide) can induce a loss of the sense of smell. It is also worth remembering that many chemicals have low effect thresholds, i.e., they cannot be detected by smell or sight until the concentration is well above the exposure limit. Others have no warning properties at all and cannot be detected by human senses at any concentration.
- Signs of spillage, e.g., stains on the floor or on protective clothing, indicate that exposure may have occurred at some point. Short-term exposures, as when cleaning up a spillage of powdered materials or cleaning down a machine with a volatile solvent, can sometimes be very high.

These simple signs can be extremely helpful, especially for an experienced person. It is often possible to assess the degree of risk on sight, without the need for any measurements, but the assessor may conclude that measurements are required to evaluate the exposure. Thus the signs should be interpreted circumspectly in the context of all the available information.

Reference should be made to any past measurements of exposure or records of effects that might assist the evaluation, such as:

- air sampling records;
- biological monitoring records;
- health surveillance records;
- illness records.

Forming conclusions and making recommendations

If possible at this stage, a judgement should be made about the level of risk, based on all the information gathered. The simplest method is to classify the risk under one of three categories, for example:

- high (urgent action needed to reduce the risk);
- medium (action desirable when resources permit);
- low (no action needed).

If it is not possible to make a judgement from the evidence available, further evaluation will be needed. This will generally involve measurement. If action is necessary, the assessor should specify exactly what issues must be addressed and suggest solutions. If it is necessary to prevent or control exposures, this should be achieved using the hierarchy of controls (see page 32).

Measurement of airborne contaminants

Levels of airborne contaminants may vary from moment to moment and from one point to another within a workplace. Therefore, the choice of sampling strategy (how to sample, for how long, where, etc.) will affect the results obtained. The strategy chosen must be appropriate to the purpose of the measurements. Several possible purposes can be identified:

- to estimate personal exposure to hazardous substances used, in order to support a chemical risk assessment;
- to demonstrate compliance with exposure limits;
- to investigate or check the performance of exposure control measures, e.g.:
  — to detect leaks;
  — to confirm that the measures continue to work effectively;
  — to verify that respiratory protective equipment is adequate.

In each case, samples may be taken as part of a specific investigation or as part of an ongoing series of measurements to monitor changes in levels.

The HSE offer considerable guidance on measurement in the form of the Methods for the Determination of Hazardous Substances ('MDHS' series) and guidance on strategy can be found in HSG173. In general, the methods adopted for monitoring gases and vapours are different from those used to monitor particulates (e.g., fumes and dusts) and standards for use in each case are specified in BS EN 482.

Sampling for particulates on personnel

The general principle of sampling for particulates involves drawing air through a filter medium at a known steady flow rate for a known time. The sample collected can be weighed and/or sent to the laboratory for analysis of active ingredients. MDHS14 provides detailed information about particulate sampling.

The specific elements of a particulate sampler are the:

- collection medium, e.g., glass-fibre filters or cellulose acetate membranes, selected according to a variety of factors including sample stability and analytical compatibility;
- sampling head, which holds the collection medium and can be designed to collect either the:
  - *inhalable dust*, i.e., the airborne matter that can enter the nose and mouth during breathing and be deposited anywhere in the respiratory system;
  - *respirable dust*, i.e., the airborne matter that, if inhaled, can reach deep inside the lung;
- sampling pump, which:
  - runs on rechargeable batteries;
  - is lightweight and portable;
  - can sample air at a flow rate up to 2.5 litres min^{-1};
  - should be capable of working at the prescribed flow rate for 12 hours.

These elements are connected to form a *sampling train*, which is calibrated prior to sampling. The sampling head is placed in the employee's breathing zone and the pump is clipped onto a belt or suitable harness. The sampling train should be checked at regular intervals to ensure that the pumps are running. At the end of the work period it should be removed and recalibrated.

Sampling of gases and vapours on personnel

In general, there are two types of sampling, *active* and *diffusive*. The general principle of active sampling is similar to that for particulates. However, the collection media are different in that they adsorb contaminants rather than filter them. There are many different adsorbents, ranging from carbon to materials specially prepared for specific compounds; the particular adsorbent used will depend on the compound to be sampled. The adsorbent is packed in tubes and the sample flow rates tend to be between 20 and 200 $cm^3 min^{-1}$.

Diffusive sampling relies on the natural process of diffusion for the collection of contaminants. Diffusive samplers are precisely engineered so as to ensure uniform collection. The samplers can take the form of badges or tubes, similar to those used for active samplers. In the majority of cases, the collected material must be analysed away from the sampling area.

Direct-reading devices

These are portable or transportable devices that can provide a direct read-out of the total amount of airborne contaminants or the amounts of specific components present. The detection principles vary from chemical detection, e.g., where a colour change occurs, to sophisticated spectrophotometric techniques. There are a few devices that can be used to measure personal exposures.

Analysis of samples

There are many different analytical techniques that can be employed, the most common being *gas chromatography* for organic vapours and *liquid chromatography* for solids.

Hierarchy of controls

The term *hierarchy* in everyday use refers to an organization with ranked levels of importance. Occupational hygienists use the term *hierarchy of controls* to mean a ranking of occupational hygiene control activities presented in the order in which they should be considered for adoption.

Controls may be ranked in order of preference as follows:

- elimination or reduction at the source;
- elimination or reduction in the path or in the environment by means of:
  — engineering controls;
  — administration and procedures;
- reduction at the employee through the use of Personal Protective Equipment (PPE).

Another way of stating this hierarchy of controls is:

- wherever possible, prevent the contaminants from being generated;
- if the contaminants are generated, keep them from being released;
- if they are released, prevent them from harming the employee.

The rationale for this ranking is provided in the following sections; it reflects the relative abilities of these controls to protect employees' health.

Substituting with less hazardous process, equipment or material

The preferred control method is to replace existing processes, equipment or materials with less hazardous ones. For example, replacing a process material with an alternative of lower toxicity reduces at source the hazard to health or even eliminates it altogether. If this is not feasible, the hazard of the system may be reduced by finding an alternative process that generates lesser quantities of airborne contaminants or requires less employee interface. However, it is not always easy to judge whether a different material is preferable and expert guidance should be sought in cases of doubt.

Engineering controls

An *engineering control* in this context may be defined as the use of hardware, other than PPE, to control the physical environment in a manner that limits the exposure of individuals to the harmful effects of a particular hazard. Examples of such engineering controls are:

- chemical fume cupboards;
- ventilation systems;

- bulk material handling equipment.

Two types of ventilation, *local* and *general*, may be used.

- *Local Exhaust Ventilation* (LEV) is the use of equipment to draw air in the immediate vicinity of a source of contamination away from those working in the area for dispersal or possible treatment elsewhere. This is an efficient way of removing contaminants from a small air space.
- *General ventilation*, also called *dilution ventilation*, is typically best used for controlling temperature but may also be used to disperse low levels of low-toxicity gases or vapours. The term covers equipment that controls the air supply and exhaust for large areas or the reliance on natural ventilation through windows, doors and other apertures in the building.

The purpose of containment systems is to segregate the employee from the hazard. Containment systems may comprise either total or partial enclosures. One example of a total containment system is where the materials being processed are confined within a glove-box. Containment systems and bulk material handling equipment are particularly useful for dealing with toxic compounds and carcinogens. They will be discussed in more detail in Chapter 7.

Administrative controls and modifications to work practices

An *administrative control* may be defined as any procedure that significantly limits daily personal exposures to chemical, physical or biological agents by controlling or manipulating the work schedule or the manner in which the work is performed. The use of PPE is not considered a means of administrative control. Administrative controls change the way employees perform the job and may include:

- job rotation;
- changing work habits;
- improving sanitation and hygiene practices.

Job rotation can be used to remedy a situation where a number of employees work in a contaminated area. It is achieved by altering working rosters so that a larger number of employees take it in turn to spend shorter periods in the affected area and thereby keep their individual exposures below established limits.

Changing work practices can be an effective way to control exposure levels, for example by:

- using a high-efficiency vacuum system instead of dry sweeping to clean up loose powder;
- adjusting the work height to avoid excessive bending;
- using material handling techniques that will prevent the generation of airborne dust.

Personal protective equipment (PPE)

Respirators, gloves, goggles and protective suits are examples of PPE. Their use is intended to prevent exposure of the human body by acting as a barrier to the harmful agents.

PPE tends to be uncomfortable and wearing it can sometimes impair efficiency as well. It can, therefore, be unpopular and is often not used correctly. Achieving adequate control of exposure using PPE is notoriously difficult.

The effectiveness of equipment designed to provide high degrees of protection will be dramatically reduced if:

- it is not worn for all of the exposure period (see page 27);
- it is inappropriately maintained or cleaned;
- the users are insufficiently trained.

The use of PPE should, therefore, be temporary, until the exposure is adequately reduced, with control at the source being the desired method of control.

In cases such as maintenance jobs, however, there may be no practicable way of controlling exposure other than with PPE.

If PPE is required, an evaluation should be performed to ensure that it is suitably selected in accordance with the findings of the risk assessment. PPE must be appropriate to both the type of hazard present and the level of exposure. It is also important to involve the work-force in the selection in order to ensure that the PPE chosen is acceptable to the users. Employees must be trained to use and maintain their PPE correctly.

Factors affecting proper selection of controls

The following factors influence which control approaches should be applied and when.

Properties of substances

The particular toxicological properties of a chemical may affect the selection of a suitable approach for controlling exposures to it. Some chemicals have no warning properties and/or very low effect thresholds. For this reason, processes or equipment using *cytotoxic* drugs, for example, would not be good candidates for the use of local or dilution ventilation to control exposure; total containment, or enclosure and isolation, would be a preferable control method.

Examples that illustrate the importance of considering physical properties in the selection process are those chemicals with very low vapour pressure, such as some oily intermediates. Since airborne exposure is typically not a concern with such materials, ventilation is not the preferred approach; instead substitution, isolation, changes to work practices or PPE should be used to control dermal exposure and ingestion.

Quantities and frequency of use (exposure pattern)

As a product becomes more widely used and increased productivity is required to meet demand, it becomes more economically viable to manufacture it using not a batch process but a continuous one with increased automation. This typically reduces manual material handling, exposure levels and potential for unplanned releases.

Installation, operation and maintenance costs

In determining which control approach makes sense, the costs of initial installation and ongoing operation and maintenance need to be evaluated against the level of protection that the various control approaches could offer. For example, providing a total enclosure around a vibratory feed bowl may reduce dust levels to within exposure limits at a relatively low cost. However, if the enclosure significantly reduces production because of the need to open the enclosure frequently to clean it or to clear blockages, then an alternative approach, for example using a mechanical or pneumatic feed, may be advisable. This would be particularly true if the alternative approach reduced maintenance costs as well as improving protection levels.

Procedures need to be implemented to ensure that all controls are operating and are being used as designed and intended. These procedures should include periodic inspections and measurements. Preventive maintenance programmes should be established to ensure that engineering controls are kept in good repair.

Employee factors

Employees' acceptance of the controls adopted is critical. Whenever possible, it is important to keep employees aware of and involved in discussions regarding changes to equipment, process and practices. They are both the people most affected and typically the ones most familiar with the hazards and variables within their jobs. Employees will endorse changes more readily if they are involved and consulted in the decision-making process.

Employees' safety and comfort are critical in process selection. All factors, not just exposure levels, must be considered. For example, if the process contains a dust that is not of high toxicity yet creates a significant slip and fall hazard when on the floor, then an enclosure approach might be advised instead of local ventilation.

Controls should be chosen so as to minimize their interference with the performance of job duties. Any control that makes job performance more difficult is more likely to be removed or bypassed if possible. For example, if an LEV system requires frequent adjustment of the position of its air intake as batch loads are brought to it, then it is not likely to be as effective as one that does not require adjustment. The use of PPE can significantly reduce productivity and work efficiency.

Conclusion

This chapter has considered some of the details associated with undertaking assessments and the options available to control employees' exposure to substances hazardous to health. The options include elimination of the hazardous substance or process, substitution with less hazardous alternatives, engineering controls including containment and ventilation, administrative controls and, finally, PPE, which should be used only when the above measures cannot achieve the required level of control.

The remaining chapters of this guide discuss ways of controlling exposure by engineering means. Further discussion of the other approaches is outside the scope of this guide.

Typical industrial operations requiring containment

4

Purpose

Chapter 1 listed some manufacturing industries in which contained transfer of hazardous materials is an important operation. The purpose of this chapter is to summarize the typical transfer operations required in such industries, since it is these that require one or more of the containment techniques described more fully in later chapters.

Contents

Introduction

Since the fine chemical and pharmaceutical process industries make their products predominantly by batch processes, the frequent transfer of materials into, out of and between items of process equipment is a common requirement. Historically, a number of processes were open and so exposed the operators to the materials they used. Nowadays the use of closed process systems is increasing and this improves the protection offered to operators. To ensure that this protection is not compromised, appropriate containment measures must be taken to protect operators and the environment from hazardous materials released during transfer operations.

Process materials involved in transfer operations can be solids, liquids, gases or combinations of these (e.g., pastes). Closed piped transfer systems from bulk tank containers can be used for larger quantities of liquids and gases. Dry solid transfers pose the greatest challenge for a containment system because of the potential for releasing fine dust and, sometimes, their poor flow characteristics.

Therefore, this guide is predominantly concerned with containment techniques for solids, and this chapter considers examples of operations in which these techniques may be needed.

Typical plant and process operations

A typical process for manufacturing chemicals by batch organic synthesis will include the following steps:

- weighing out the solid raw materials;
- adding solid, liquid and/or gaseous reagents to reactors and process vessels;
- reactions;
- one or more purification steps, which could include addition or removal of solids;
- crystallization of the solid product;
- separating product solids from the mother liquors;
- drying;
- removing the product into containers;
- milling and blending.

Large-scale plants involving tonnes of solid materials per batch may incorporate dedicated hoppers with closed mechanical or pneumatic handling systems. These may be used for both raw materials and final products. They provide good intrinsic containment but the materials are unlikely to be particularly toxic except in a few cases.

Within such plants, typical transfer operations for which containment may be required include the following:

- solids:
  — sampling of sacks or drums, e.g., during transfers to containers, semi-bulk bins or process equipment;
  — tipping drums or sacks into hoppers and vessels;
  — transfers from one drum to another for weight adjustment;
  — transfers between items of process equipment, e.g.:

— granulators to blenders;
— mills to blenders;
— blenders to tabletting machines;
— breaking connections after emptying semi-bulk solid and flexible containers into process vessels;
— discharging centrifuges and filters or cakes into containers;
— charging and emptying open dryers;
— filling drums from dryers and blenders;
— filling semi-bulk containers from dryers;
— transfers from semi-bulk containers to blenders, mills, etc.;
— final product packaging, e.g., to bottle fillers or blister packing machines;
• liquids/vapours:
  — line breakage from coupling or decoupling of bulk tanker;
  — emptying, filling and sampling drums;
  — sampling reactor vessels and lines;
  — coupling or decoupling pressurized liquid containers;
  — openly handling solid cakes wet with solvent;
  — filling small containers with final products;
• gases:
  — coupling/decoupling of pressurized gas cylinders.

Typical operations requiring containment

Typical process operations requiring containment include:

• milling and micronizing;
• blending;
• granulation and coating;
• tabletting;
• filling bottles and foil packs;
• dispensing;
• filling or emptying tote bins.

Plant operations – batch reactor plants

The general mode of operation for a typical batch reactor plant is shown schematically in Figure 4.1, where the transfer operations are marked by asterisks (*). The use of Intermediate Bulk Containers (IBCs) of typically 1000-litre capacity is indicated as an example of a container system for transferring solids.

At one or more points in the process, samples may be taken, particularly from the reactor, from the separation and drying equipment and from the final product container(s).

Some products may be packaged into sales containers in the production plant. Others may be transferred from the production plant to a formulation and final packaging plant in differently sized containers ranging from 15 kg drums to semi-bulk flexible or rigid bins containing 200 kg or more. Liquid products may be poured or piped into drums for transfer to

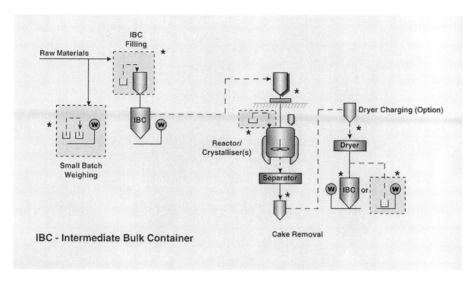

IBC - Intermediate Bulk Container

Figure 4.1 Organic synthesis plant

the formulation and packaging plant or transferred within the production plant directly into smaller containers for sale to the customer.

Process operations – formulation and packaging plants

Formulation and packaging plants rely on a range of different operations, but all require material transfers into and out of the process system. Process operations are dependent primarily on the product type, the final customer or (for a pharmaceutical) the therapeutic formulation requirement.

Pharmaceuticals in solid dose form and other blended products may require a sequential range of different unit operations involving weighing, blending, milling, granulation, drying, tabletting or encapsulating, coating and packaging into bottles or blister packs. In these predominantly dry process operations there is the potential for release of hazardous materials during transfers between items of equipment and some in-process operations.

Figure 4.2 (page 40) illustrates a generalized process for manufacturing pharmaceuticals in solid dose form, with the transfer operations marked, as before, by asterisk*.

Whilst traditional plants have utilized discrete unit operations with containerized movements of process materials between stages, as Figure 4.2 suggests, this approach can involve significant handling and transfers. In order to avoid this, the current trend is to couple unit operations closely together, with gravity and/or vacuum transfers of solids between them, and so eliminate open transfers.

Sterility and containment

Parenteral pharmaceuticals are intended to be injected into the body. If it is to be done safely, they must show no harmful microbial or pyrogenic activity. They must, therefore, be produced

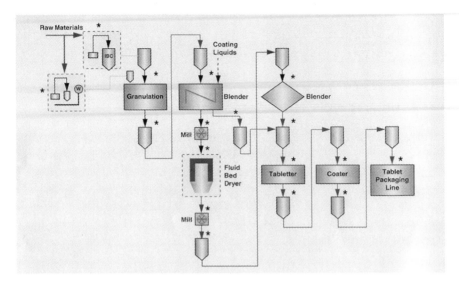

Figure 4.2 Solid dosage plant

under sterile conditions, which means that they, and any gases, solids and liquids included as a part of the product, must be kept free of microbes and pyrogens. Contact between operators, the environment and the product must be severely limited or, wherever possible, avoided altogether. The containment philosophy traditionally adopted has been to protect the product from biological contamination from the operator rather than the other way round and the arrangements adopted to achieve it could include sterile air flows away from the product towards the operator, all within clean-air environments. Traditional industrial practices are well documented and this guide does not attempt to repeat the containment techniques used specifically for protecting sterile products.

However, the increasing use of more potent drugs in sterile delivery forms necessitates containment measures to protect the operator as well as the product. This guide covers the principles of isolation containment barriers around the process, but is not intended to cover the many specific additional details which sterile process operations impose on the isolator and its support systems.

Plant size and transfer quantities

Plant sizes and their outputs vary enormously across the fine chemical and pharmaceutical industry, as do the harmful effects of the materials they handle and produce. A few products, intermediates and agrochemicals may be made in a few thousands of tonnes per year in each plant. Some of these products, e.g., weed-killers, may be very toxic. Some active pharmaceutical ingredients, such as pain-killers, are also made in bulk in plants of this capacity range in order to achieve economy of scale, but they have very low potencies.

Most fine chemicals and active pharmaceutical ingredients are made in plants with annual capacities ranging from hundreds of kilograms to thousands of tonnes. There is, however, an increasing number of pharmaceuticals with potencies that require relatively small patient-dose quantities. Pharmaceuticals may also be focused on diseases that affect only a very small percentage of the population; plant outputs for these may be very small, ranging from 50 kg to 5 tonnes per annum.

The quantities of hazardous material transferred in discrete amounts depend primarily on the batch size, which in turn depends on the reactor (and crystallizer) volumes used and the size and type of product container. These are summarized in a generalized form for the fine chemical and active pharmaceutical ingredient industries in Table 4.1.

Table 4.1 Typical container transfer sizes

	Typical material quantity per transfer, kg				
Transfer operation	**<1**	**1–25**	**25–50**	**50–300**	**150–1000**
Sampling	*				
Drum/sack tipping		*	*	*	
Drum weight adjustment	*	*	*	*	
IBC filling/emptying — synthesis plant				*	*
Pilot plant transfers		*	*		
Transfers of very toxic material	*	*	*		

As plant outputs increase and batch equipment sizes increase accordingly, it is typically the number of transfers that goes up per batch (limited by container size) and not the size of the container used for the transfer.

The quantities of toxic reagent liquids and gases to be transferred into the reactor are a function of the process chemistry and reactor batch size. For liquid reagents, they typically range from 5% to 20% of the reactor volume.

In production plants involving a continuous campaign of sequential batches, the frequency of transfer operations will depend on the batch cycle time, but will typically be once every 10–15 hours. In research and development pilot plants making one batch at a time, transfers may occur once every few days or even less frequently.

Transfer operations involving solids

Solids charged to processes include:

- organic chemicals:
  - basic building blocks;
  - intermediates;
  - final products;
- inorganic chemicals, e.g.:
  - aluminium chloride;
  - carbonates;
  - chemical salts;

— cyanides;

— metals, e.g., zinc;

- small quantities of 'inert' solids, e.g.:
  — filter aid;
  — activated carbon;
- catalysts.

The transfer techniques used depend on the choice of the containers used to transfer solids into or between stages of the process and to hold product solids at the end of the process. This choice is in turn dictated by scale, product toxicity and end-user requirements. Container types range from:

- small (5–10 kg) containers used for:
  — manual transfers; or
  — transfers of very toxic materials;
- drums or bags holding 15–50 kg, usually with plastic liners, for which manual or power-assisted transfers are used;
- semi-bulk flexible or rigid containers holding 100–1000 kg of material, with transfer effected by gravity discharge to the process system;
- hoppers and silos holding tonnes of material.

The transfer techniques used predominantly are:

- gravity flow from one container to another or to the process vessel;
- pneumatic conveying using air or nitrogen as the scale and frequency of operation increase.

Simple gravity transfers are preferred for materials of higher toxicity.

Dispensing operations

The off-line weighing out of the quantities of solids or liquids required for a process batch is termed *dispensing* or *batch splitting*. The materials to be dispensed are either raw materials bought into the plant or intermediates made by previous process steps. The transfer operations required include:

- manual transfers of small quantities of material from one small container to another to meet the desired charge weight;
- transfer of the desired weight of material from drums or bags to semi-bulk containers, using techniques such as:
  — manual or power-assisted tipping;
  — vacuum transfer;
  — pneumatic transfer;
- transfers from one semi-bulk container to another using gravity or pneumatic or vacuum arrangements.

Charging of vessels and equipment

Solids are charged to vessels either to make up solutions or as reagents or additives for reactions. Transfer operations include:

- manual or powered open tipping of dispensed containers, either via the man-way or via a dedicated solids charge port;
- gravity discharge of material from a semi-bulk container or a hopper located above the vessel.

In some cases, a rotary valve or screw feeder is used to measure out accurate quantities and to ensure that the product cannot escape.

Unloading solid-liquid separation equipment

In most plants, product cakes from filters or centrifuges are discharged directly to a dryer. These cakes will normally be wet with solvent. However, in some processes, wet cake may be off-loaded from the separator into drums, mobile skips or bulk bins and charged to a process vessel or to the dryer as a separate operation.

Unloading dryers

Solids are usually transferred by gravity directly from the batch dryer to the final container, either to a nominal weight or via a feeder system to an exact weight. Where continuous dryers are used on larger batch processes, one of two approaches may be adopted:

- two filling heads may be used alternately to fill one container while another full one is removed or an empty one inserted in its place;
- solids may be discharged into a buffer hopper, which is emptied periodically.

Pneumatic transfers may be used to transfer materials from the dryer to the offloading hoppers. On the small scale, some types of dryers, e.g., tray dryers, may require manual unloading of their contents into containers.

Milling and blending

After drying, it may be necessary to mill the product to a uniform particle size and then blend the batch for consistency. This operation may be 'in-line' after the dryer, with a direct, closed connection to the mill, or 'off-line' where separate charge and discharge transfers to the mill and blender will be required.

Removal of solid impurities

During batch processing, filtration may be required to remove particulate impurities or other solids from the process. These impurities will then have to be removed from the filter and, since they may be toxic or may release solvent or chemical vapours, containment measures are required for this process. It is quite common to wash these solids with water to remove solvents before discharge.

43

Transfer operations involving liquids

The range of liquids requiring transfers into chemical processes includes acids, alkalis, solvents and chemical reagents. These in turn may include liquids with boiling points at or below ambient temperatures and consequently held under pressure. Liquids transferred from chemical or formulation processes include aqueous wastes, waste solvents and liquid products, either in a pure form or as a diluted solution.

Liquid transfers are obviously easier to contain than operations involving solids.

The hazards associated with open liquid transfer operations include:

- operator contact with splashes;
- operator contact with toxic fumes;
- release of uncontrollable spills to inappropriate drainage systems;
- release of uncontrollable vapour flows from pressurized containers to the environment.

The following sub-sections summarize those instances where open transfers and enforced breaks in closed piping systems commonly occur.

Loading and unloading of road tankers

Flexible hoses with proprietary "dry-break' couplings are used to connect the tanker discharge to the filling couplings on the piping system associated with a storage tank. The major hazards they present are:

- a catastrophic failure of the joint during transfer, causing the contents of the tanker to escape;
- the possibility that the tanker may depart with the hose still connected.

It is also important to consider how the gases and vapours displaced from the receiving container will be dealt with. Where these present no safety or environmental hazard they may be vented to the atmosphere but otherwise separate vessels and associated piping must be provided to contain them, for example by returning them to the source container. In either case, scrubbing, condensation or equivalent technology may also be needed.

After the transfer to or from the tanker has been completed, residues of the liquid may remain in the hose and drip out. Any such leakage must be minimized and the spilt liquid properly contained.

Drum and semi-bulk container transfers

The transfer of chemicals, some of which could be very toxic, from drums into process systems is a very common operation. Transfers are usually achieved by applying vacuum or self-priming pumps to the liquid via a dip pipe lowered into the drum. The main considerations to be addressed here are:

- containment of vapours from the open drum;
- drips from the dip pipe after removal;
- decontamination of drums, ideally carried out *in situ* after use.

Small container transfers

Small charge quantities of 10–20 litres may need to be transferred from drums to portable small containers. The portable containers must then be coupled to process vessels and the contents transferred by vacuum or gravity. This may be achieved in a number of ways but all require direct coupling of the container nozzle to a piping connection local to the equipment. A number of proprietary devices are available for this.

Pressurized liquid transfers from cylinders

Pressurized liquids are usually transferred from cylinders into processes by coupling the discharge nozzle of the cylinder to an equivalent coupling on the end of a flexible hose attached to the rigid piping system. Various purging protocols may be required before and after coupling and decoupling and containers may be held in the open air or in segregated closed rooms with scrubber systems.

Container filling

200-litre drums or smaller containers at atmospheric pressure may be filled using proprietary filling devices. The main containment problem is to capture the vapours displaced.

Pressurized container filling is obviously achieved in closed systems. Making and breaking the couplings are the most hazardous aspects of the transfer operation.

Gaseous transfers

The methods of transferring gases to process systems are similar to those for pressurized liquids. Again, making and breaking the couplings are the most hazardous operations and their engineering integrity should be very high. Purging of the coupling pipework before removal is the usual practice.

Sampling

Small solid samples are often taken from:

- raw material containers, to check the quality of the material against the identification on the container and the supplier's documentation accompanying it;
- the separator and/or dryer, to check for residual solvent;
- one or more of the final product container(s).

Liquid samples often need to be taken from process or reaction systems since, to date, the use of on-line chemical analysers is limited. Such sampling is particularly important in the case of toxic solutions from reactors and other critical process steps. Traditionally, sampling was achieved by manually inserting a pipette into the batch via an open man-way. Permanent dip pipes and vacuum or recirculation systems with proprietary sampling devices are now available, which offer a more contained approach and, for solvent processing, allow an inert atmosphere to be maintained in the vessel.

Process containment

In chemical reaction processes, devices are employed to relieve pressure surges in closed process systems. Such devices include relief valves, bursting discs and containment systems such as scrubbers. Very large filters will be required to control the emission of hazardous materials via the vent gases from the catch tank. This guide does not attempt to discuss these traditional protective features of closed systems, which are adequately covered in the respective manufacturers' literature.

Plant cleaning

Process plants, particularly those operating as multi-product facilities, have to be cleaned between campaigns. Whilst *Cleaning-in-Place* (CIP) techniques are increasingly being used, it is often necessary to dismantle equipment containing harmful process residues for manual cleaning. When this occurs, appropriate measures must be applied to ensure that the chemicals are contained and personnel protected.

Conclusion

This chapter has identified some typical industrial operations that have the potential to expose employees to harmful concentrations of hazardous materials and therefore require provisions to contain these materials. The next chapter introduces the principles according to which such provisions can be implemented. Practical applications of those principles will be discussed in further detail in Chapters 6 and 7.

Principles for the design of containment systems

5

Purpose

The purpose of this chapter is to introduce the principles of containment system design and consider their impact on the overall plant. Since the selection of an appropriate containment system is based on the properties of the materials handled and the type and scale of operations, a broad classification system for these is identified.

This chapter, therefore, provides the basic information needed to select a suitable containment system and acts as an introduction to later chapters, in which the selection, design and performance of containment equipment are discussed in more detail.

Contents

Historical perspective

Figure 5.1 shows how the approaches adopted towards containment of hazardous substances have developed and become more sophisticated. A four-step historical approach to containment systems can be identified.

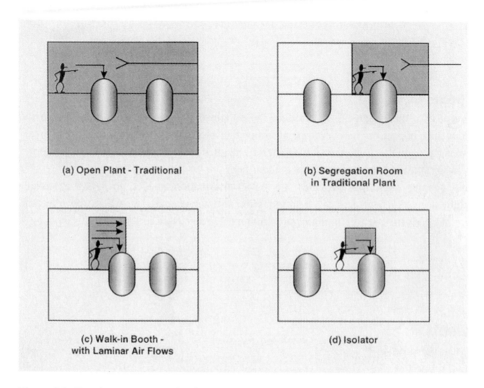

(a) Open Plant - Traditional

(b) Segregation Room in Traditional Plant

(c) Walk-in Booth - with Laminar Air Flows

(d) Isolator

Figure 5.1 Containment systems development

The simplest approach, as shown in Figure 5.1(a), was to surround the hazardous material with a Local Exhaust Ventilation (LEV) system, possibly backed up by containment systems such as air-locks, room filters or scrubbers, assume that the plant within which hazardous materials are exposed in open transfers is totally contaminated and require the operator to wear Personal Protective Equipment (PPE) before entering the area.

The next development, as shown in Figure 5.1(b), was to locate the transfer operation in a dedicated room to prevent total contamination of the plant.

A further general improvement was the construction of a closed booth in the plant, as in Figure 5.1(c), in which the open transfer operation would take place. Such a booth would be fitted with a local extraction system and would restrict the potential spread of dust or vapour to the adjacent space and so avoid contaminating the rest of the plant. Where manual operations, such as tipping, were required, the operator would carry these out in the booth, wearing PPE as a form of protection against the locally contained contaminant. This approach could still lead to the spread of contamination onto the operator's PPE, where it could be carried away from the transfer zone via the operator, and so arrangements for washing the PPE were required adjacent to the booth. This arrangement is still considered unsatisfactory as it relies on PPE rather than engineering controls to protect the work-force.

Advances on these systems incorporate improved air-flow containment systems in the booth to control the spread of dust or vapour during open handling.

The current approach adopted for more toxic materials aims to use a closed system, or *isolator*, as shown in Figure 5.1(d), to prevent the spread of hazardous material and avoid the need for PPE.

Material properties

All transfer operations involving hazardous materials present a hazard to the operator. A suitable containment system must, therefore, be chosen that will reduce the level of exposure to below a safe, acceptable limit.

It is commonly accepted that the degree to which operators may be exposed to and harmed by the release of hazardous materials in transfer operations depends on four main factors, namely:

- the physical form of the material;
- the scale of the operation;
- the hazardous properties of the material;
- the methods used to handle the material.

The choice of containment system for an operation depends on the overall hazard represented by these factors. It is therefore necessary to categorize these factors; methodologies for doing so and for identifying appropriate containment strategies accordingly are described in Chapter 6.

The exposure could be the result of direct contact with:

- the process material during transfer;
- residuals or releases from mechanical or closed transfer systems during plant cleaning and maintenance operations.

The duration of potential exposure is important for occupational hygienists, since operators' exposure limits are generally stated as time-weighted averages. For example, if a dispensing operation is carried out for 10 minutes in every shift, that is clearly a much shorter time than the 8 hours across which the long-term exposure limit is averaged. It can, therefore, sometimes be permissible for the operator performing the transfer to be exposed to a higher concentration than that quoted as the exposure limit, provided the exposure averaged over the whole shift is below the long-term exposure limit. Where a short-term exposure limit is also specified, exposures averaged over any 15-minute period must not exceed this level. However, it is generally recommended that exposures should never exceed 3 times the exposure limit over a short period.

A rule of thumb sometimes used for containment system design is that the expected concentrations at the operator's breathing zone should be $\frac{1}{10}$ of the allowable exposure limit. This design margin accommodates variations in equipment performance and operating practices.

Another point to note is that operators are exposed to dusts in their normal breathing zones. Concentration here is not necessarily the same as the calculated or measured average particulate concentration in the room: if it is lower, there is no problem, but in general it is likely to be higher and increase the operator's exposure. Ideally, the operator should be positioned such that the breathing zone is away from the dusty zone around the point of transfer.

Operators can be protected from process materials by the use of PPE such as air-suits, or by engineered solutions that provide barriers of some form or another around the hazardous material. However, UK health and safety legislation requires that the control of hazardous material release should be achieved primarily by appropriate engineering designs and operating techniques, in accordance with the hierarchy of controls described in Chapter 3. PPE should be used only:

- for recovery after a fault;
- as a back-up measure, where the engineering solution cannot totally achieve the level of protection required;
- for infrequent use where the cost of an engineered solution cannot be reasonably justified.

Corporate policy

To comply with the requirements of the **Health and Safety at Work etc. Act 1974**, many companies have a clearly defined corporate safety policy. Issues covered in such a policy should include:

- the need to assess the risks from materials handled and their toxic properties;
- the need to provide Material Safety Data Sheets (MSDSs) for each material handled;
- the need to identify appropriate containment equipment and develop operating procedures to meet the hazards of the materials handled;
- emphasis on engineering containment and administrative controls rather than PPE;
- regular monitoring of operating environments and operators to determine exposure levels;
- a statement on the categories of employees who should be involved in assessing safety and system design features, e.g.:
  — management;
  — corporate safety and hygiene personnel;
  — operators;
  — other employees or other persons who may be affected.

Plant operations and integrated process solutions

Suitable containment systems may have to be fitted retrospectively to:

- existing plant for an existing product;

- existing plant adapted for a new product of increased risk;
- new facilities, in which case the containment systems should ideally be developed as part of the integrated design of the new facility.

In all these cases, a containment problem may be associated with one unit operation, though it is more likely to concern a range of unit operations across the plant from reactor charging to dryer pack-off. For example, the container filled with an intermediate from a dryer may also be the container used to charge a reactor. Traditionally this has been a flexible, plastic- lined drum, which does not provide good containment when the products are emptied from it. Consequently, for existing and new plants, it is best to consider all transfer operations in Chapter 4 together and develop an integrated plant solution.

In addition to a consistent containment solution within the plant, there may also be a need to consider the overall materials supply chain. Suppliers may be requested to deliver hazardous raw materials in the container designated for use with the plant's containment system. More importantly, it may be advantageous to use similar product containers and handling systems in both the bulk plant and its downstream formulation plant.

On existing plants, open processing equipment and transfer operations may require a high usage of PPE. It may be difficult to achieve the improvements that would minimize release to such an extent that PPE usage can be eliminated, unless significant changes are made to introduce closed processing equipment. Since this may not be economically justifiable, only marginal improvements may be practicable. However if the spread of hazardous materials within the plant can be noticeably reduced, and the overall plant safety levels improved, then the containment project is probably worthwhile.

Back-up containment systems

In addition to the principal containment equipment and other systems provided in a plant, back-up containment systems, usually around a plant operating area, may be required. Their purpose is to protect the rest of the plant or the environment either:

- if the principal or subsidiary system fails; or
- to capture the small releases into the operating area from the principal containment system.

Protective systems typically include:

- a scrubber fitted to a room ventilation extraction system to remove stray or accidental vapour releases from transfer operations within the room;
- *High-efficiency Particulate Arrestor* (HEPA) filtration fitted to room ventilation exhaust streams to capture stray dusts not removed by the principal containment equipment;
- pressurized air-locks at room entry/exit points;
- water showers to decontaminate operators (in their PPE) and containers before they leave the operating area;
- containment sumps with filtration, chemical or other treatment methods to collect and possibly detoxify spillages.

The designer and owner of the plant must judge the extent to which these containment measures are required. Questions they should ask include:

- what are the probability and likely scale of a leakage from the selected containment system?
- would the harmful effects on the environment or external personnel be too great if a defined or uncontrolled leakage occurred and no back-up systems were installed?
- should the back-up containment system be applied locally to the potential release point rather than totally around the room or plant area?

Containment system design principles

The basic principles for containment system design are:

- substitution of material, e.g., to replace benzene (which is toxic) with a less hazardous alternative;
- substitution of form, e.g., to replace a fine powder with less dusty pellets;
- to analyse and change the process to minimize or avoid transfer steps and particularly open transfers;
- to use closed transfer process techniques where possible to minimize the contamination zone around the point of release;
- to analyse all steps of the process involving transfers in order to identify a common integrated containment solution.

The practicality of a closed system must be assessed with regard to:

- scale;
- properties of materials used in the process;
- containers that have to be used.

The extent of capital expenditure for a containment system, and particularly a relatively high-cost closed system, must be based on:

- the assessment of risk;
- corporate policy;
- availability of a system to meet the process operation;
- economic justification, particularly for retrofits on existing plants.

The selection of a suitable containment strategy for the materials used and scale of operations is discussed in Chapters 6 and 7.

Regulatory considerations

Regulatory issues for containment system design

The design, installation and operation of pharmaceutical and food plants must meet the current requirements of *Good Manufacturing Practice* (cGMP). These requirements, defined in US

Federal Regulations and equivalent EU Directives (actual or proposed) and national legislation of EU member states, are most stringent for pharmaceutical plants, particularly those that manufacture products in their final dosage form. As has been shown earlier, a containment system in a pharmaceutical plant not only protects plant operators and the environment from the active process material but also protects products from contact with any potential contaminants in the environment. The latter is a primary aim of cGMP. Both are safety issues affecting humans.

The key questions to ask when applying cGMP principles to a containment system in a pharmaceutical plant are:

- will the equipment, or its components, in which a drug product is being processed contaminate the product?
- could one drug product be contaminated by another product which is being made either:
  — in other equipment at the same time; or
  — in the same containment equipment sequentially?
- is the material being processed the correct material for that batch and charged to the correct batch weight?
- are the processing conditions and operating procedures always the same, and as prescribed, for every batch of the same material?

Selection of equipment and materials

Containment equipment must not produce particulates or chemicals that could come into contact with and so adulterate the products being processed. This is especially important when final products or their equivalent chemical forms are being produced. In general, surfaces must not shed particulates or leach out chemicals; this requirement particularly applies to elastomeric seals and components in contact with liquids or subject to solid abrasion.

HEPAs or other filters in close proximity to products in isolators must not generate particulates in operation or be affected by solvents or chemicals used for cleaning. All filter components that are or could be in contact with the product (e.g., seal fluids or air exhausts) should be analysed for conformity and not cause particulate, chemical or other contamination.

Cross-contamination

Cross-contamination is a significant issue for pharmaceutical plants that produce more than one product or intermediate at the same time or in sequence.

In order to avoid the potential for cross-contamination during parallel processing, the following principles should be applied in the design of containment systems:

- recycled air should not carry over material from one containment system to another;
- components that come into contact with the product should not be used for more than one product stream at a time;
- if air-flow containment booths are located adjacent to each other in the same room, no cross-over contamination from air-flows, open empty containers or operators should be allowed to occur;

- operators handling different materials by open transfer in different booths or on different equipment should use separate changing facilities to avoid cross-over contamination from their clothing;
- physical, electronic or procedural checks should be applied at each transfer point to ensure that the correct product container is emptied or filled to the designated batch weight;
- shared plant should be cleaned, using a validated cleaning system and procedure, between use for different products;
- if containers are moved around the plant (or external areas) they may pick up undesirable contamination, both particulate and chemical, which could be taken into the product process system; their coupling systems may need protection against this possibility;
- where closed system connecting points for different products are located adjacent to each other, care should be taken to avoid cross-over contamination from container movements, people or air-flows.

Plant cleaning

The requirements for cleaning containment systems, particularly isolators, should not be underestimated. The use of more sophisticated engineering in closed systems for the more hazardous materials requires that more rigorous cleaning routines should be followed before the systems are opened up or used on another product, and yet this complex equipment can be intrinsically more difficult to clean.

Cross-contamination from residues left in equipment at product change must be avoided and equipment designs must ensure that cleaning processes are efficient and reproducible. Appropriate, validated cleaning procedures are required to protect both the products and personnel from contamination by hazardous materials whenever the equipment is opened up for maintenance. The following design principles should be applied with regard to cleaning of containment systems:

- equipment that comes into contact with the product should be easily cleaned and should therefore be designed with:
  — rounded corners;
  — no crevices;
  — polished surfaces;
  — no 'shadows' where sprays used for Cleaning in Place (CIP) cannot reach and which cannot be manually wiped;
  — drainage that removes all liquid with no pooling;
- allowances should be made for swab testing after cleaning where necessary;
- those components that have to be changed, e.g., filters, should be easily accessible and removable, usually whilst contaminated, in a safe manner;
- the ergonomics of operator reach and physical capability when cleaning booths and isolators manually should be evaluated;
- how the cleaning fluids are supplied to the containment system and drained out of it should be considered;
- the benefits of a permanent CIP system should be evaluated;

- seals and instruments that come into contact with the product should be hygienically designed or else eliminated;
- operators who need to dismantle equipment that may be contaminated should be provided as appropriate with:
  — PPE;
  — showers;
  — clean changing areas;
- all cleaning procedures to be applied during use or maintenance of the equipment should ideally be validatable, i.e.:
  — defined;
  — reproducible;
  — verifiable;
- if one or more validatable procedures cannot be developed, then swab testing after individual clean-outs will be required;
- if the equipment needs to be dismantled for cleaning, the procedures above must include detoxification where necessary;
- the effects of cleaning fluids on the materials of construction must be considered, as discussed further in Chapter 7, when procedures are drawn up and materials selected for cleaning; in particular, the effects of solvents on elastomeric seals and filters must be noted;
- cleaning facilities for containers (e.g., Intermediate Bulk Containers (IBCs) and semi-bulk containers) should be located and operated so that the waste materials removed can be disposed of safely and clean containers do not become contaminated;
- maintainable items that do not come into contact with the product should, wherever possible, be located outside the containment or isolation equipment, so that technicians can carry out their tasks without the need to decontaminate inside the container first or wear PPE.

Material identification

The containment system must be designed and operated so as to ensure that closed containers holding process materials are correctly tagged and clearly identifiable by material type and quantity at the point of use. Any errors in the identification of materials compromise both cGMP and process safety implications. The checking systems typically in use include:

- visual and electronic tags on small containers and IBCs, with identification readers at the point of use;
- weighing scales inside booths, isolator cabinets or under IBC fill and discharge stations to measure or check weights;
- actuated valves on charging and filling systems, which open only when the control system recognises that the correct container is in place.

Repeatable process conditions

This requirement relates mainly to the use of defined operating instructions and the calibration of process control equipment inside or associated with containment systems. For materials handling containment systems, this usually involves weighing instruments and sampling devices.

Where scales are located within isolation systems, the design should allow access for calibration without the need to dismantle the system. The design should also allow access, for maintenance and removal, to other instruments located within or penetrating the containment barrier without extensive dismantling or loss of the containment seal.

Various devices are available for the removal of small samples of liquid and solid materials from booths or closed systems. The process designer must consider the ergonomics of their operation and their compatibility with the overall containment standard required.

Plant layout considerations for contained operations

Whilst process and regulatory considerations generally dictate equipment arrangements and the plant layout, consideration should also be given to transfer operations and their potential for spreading hazardous materials around the plant if they are not contained. The typical solid and liquid transfer operations for the batch chemical and formulation industry have been identified in Chapter 4.

One key layout consideration relates to the hazard posed by the material if or when it is exposed. For extremely toxic materials (equivalent to bands E and F as described in Chapter 6), it is usually recommended that dedicated plants or processing areas with suitable back-up protective systems, such as negative pressure plant areas with air-locks or plant exhaust ventilation systems with HEPA filters, should be considered.

Layout considerations – solid containment systems

Where close-coupled transfer operations take place between items of equipment, the basic layout configuration must meet the needs of the specific process used. For example:

- materials should normally be allowed to flow under gravity;
- it must be possible to:
  — access the equipment for maintenance;
  — dismantle it to remove contaminants during cleaning;
- areas of potential exposure to different materials must be segregated.

If materials from the process could contaminate the local plant area, then segregation of the plant area and its drainage and air filtration system from the rest of the plant should be considered.

When close-coupled IBCs are used for transfers, good access must be allowed for the operator to handle the containers with powered or manual hand trucks. For plants where many movements of discrete containers take place, an automated container handling system (relying on an *Automatic Guided Vehicle* (AGV) or mechanical handling) may be cost-effective. In this case, the overall plant layout is usually controlled by the demands of the handling system, including access to and from the local materials storage or holding areas.

A separate floor above the chemical processing equipment is often used for charging the process from fully closed IBCs or equivalent-sized containers or from drums in a tipping booth. This provides two advantages:

- if there is a spillage the plant itself is not contaminated;
- more space can be made available for IBC movements.

If a fully sealed transfer coupling system is used, these areas may be built to a lower specification, i.e., that of a *technical area*, than the production areas below in order to reduce cost, provided they can be cleaned easily.

In existing plants without an upper charge floor, solids should be charged directly to the reactor or process vessel. A range of containment devices is available for such operations, ranging from unidirectional flow booths through to isolators closely fitted to the process vessels.

Where IBCs and other containers are manually filled, this operation is best kept away from complex chemical plant processing areas to minimize the spread of dust. It is usual, therefore, to attempt to segregate dispensary, charging and pack-off operations from other process equipment and piping areas.

Where exposed materials are handled, hygienic and cGMP design considerations will require the containment systems to provide an easily cleaned, non- contaminating environment and to segregate one product from another in multi- product plants.

Containment devices, such as isolators, are usually located within normal plant areas or dryer pack-off rooms where they can be close-coupled to process equipment for transfers to and from the process. Where larger batch quantities or many discrete loads are handled within containment isolators, powered, remote-handling systems may be employed. Space must be allowed, in plants with a high throughput, for suitable conveyors to move containers into and out of the isolator.

Layout considerations – liquid containment systems

Liquid transfer operations requiring containment are summarized in Chapter 4.

The basic principle of layout design is to minimize the spread of hazardous materials. Drum- or cylinder-emptying booths, in which liquids are transferred by pipe to process equipment or smaller containers, may be located in the open plant if they are provided with unidirectional air-flow to protect the operator. If the materials handled are considered very toxic, then the booths would be located within a smaller closed area with suitable back-up containment systems (e.g., space scrubbers with closed drainage) in case of leakage. The coupling points that enable small transfers of toxic materials from portable containers to a process system are normally located within cubicles in the plant as close to the process system as possible.

For the most toxic materials, drums should be opened inside closed isolators. At the extreme, automated, powered remote handling devices should be used. Facilities may also be provided for detoxifying containers at the point of use, although central washing facilities are also common in larger plants.

For hazardous liquid products, container-filling points should be located in containment booths, which may be located in the general plant area or in a smaller 'closed' room. For very toxic materials the latter option should be used, with additional back-up protection.

Integrated filling lines must be installed in such a way as to provide a suitable containment strategy at the point of fill and before the container is closed.

Aerosols, if generated, should be removed by LEV points or unidirectional air-flows.

Ergonomic issues

The designer of manual handling systems must make allowances for the reach, height, body shape and load-carrying capabilities of the actual operator or group of operators carrying out the tasks in their actual operating positions. While this may sound obvious, these ergonomic issues may easily be overlooked during the design process and so increase exposures, restrict the expected performance or physically harm the operator.

Some examples of ergonomic constraints with manual systems are as follows:

- the height from the floor of a tip station for unloading drums or bags will depend on the operator's height;
- the exposure of an operator to dusts during manual scooping of material out of a drum will depend on the depth of the container and length of the operator's arm;
- the maximum dimensions between the activity point(s) and hand positions inside an isolator fitted with glove-ports will depend on the length of the operator's arm;
- operation of drum wands and valves inside ventilated booths cannot be beyond the reach of the operator;
- the size of isolator gloves and half-suits may have to be selected for the specific operator carrying out the tasks;
- the position of regularly maintained items inside isolators may require glove-ports to be located specifically for these operations.

Isolation systems present special ergonomic challenges. Their shape and location in the plant depend on the tasks the operator has to carry out. In order to produce a satisfactory design the following steps should be taken:

- collection of data affecting the ergonomics of the design:
  — exact operations to be carried out for both process and maintenance;
  — characteristics of the materials to be handled;
  — equipment characteristics of the proposed isolation system;
  — user group characteristics;
  — materials to be transferred to/from the isolator;
  — plant constraints;
  — cleaning/dismantling needs;
- preparation of preliminary design;
- construction of a cardboard/wood mock-up to test all operator and material interfaces;
- preparation of the detailed design based on modifications to the proven mock-up.

The most important consideration to note is that an isolator severely restricts the physical operations involved in handling materials under both 'normal' and 'safe operator' procedures. Factors to consider include:

- reach;
- handling of weights at various distances from the body;
- the potential for weights or moving equipment to injure the operator;
- equipment modifications to allow for limited access;
- frequency of usage;
- operator comfort.

All of these concerns should be covered by attention to all manual and other operations to be carried out and the thorough testing of prototypes that exactly model the operational plant.

Process system design for maximum intrinsic containment

The need was identified on page 51 for back-up physical or chemical containment systems where high-risk materials are handled and the principal containment device could fail or be compromised by a malfunction of the process or operator error. Obviously, the costs associated with these systems can be minimized if the hazard is eliminated altogether or if the principal containment system is engineered so as to eliminate or significantly reduce the chance of escape. All system designers should ensure that:

- consistent design standards and ratings are applied throughout;
- there are no weak links in the system;
- closed systems are used wherever possible.

Weak links in the system commonly occur in the coupling of components to containers. Some examples include:

- liquids:
  - low-pressure push-on hoses with jubilee clips (not recommended);
  - handling from open containers;
  - open drips from couplings with no flushing devices;
  - low-pressure plastic or glass containers coupled to 6-bar reactors;
- solids:
  - canvas or thin rubber connection pieces;
  - unprotected flexible sacks or other containers holding toxic material;
  - open tipping or manual handling;
  - thin-walled low-pressure containers or hoppers coupled to 6-bar reactors;
  - lack of pressure-rated interlocked devices to ensure safe transfer of materials between the process and 'weak' containers.

The expediency of using very simple, easily handled devices must be balanced by the risk of escape and the impact of released materials on the operators and the surrounding environment. It must be considered whether relief devices or interlocked pressure-rated valves, etc., should be installed between the process and weak containers coupled to it.

When the containment component is part of the closed system, consistent pressure-rated devices should be used, such as reinforced hoses with rated coupling devices.

In addition, it must be possible to clean inside the closed system to remove toxic materials from internal parts before uncoupling.

Intrinsic process safety is also improved if the integrated process system approach introduced on page 50 is applied. Eliminating transfer steps and transient containers will improve process safety by minimizing risk points, through the provision of better-quality engineered containment systems.

Conclusion

This chapter has identified various techniques that can be adopted and questions that should be considered during the design of containment systems. It has also indicated that some techniques are acceptable only for materials of relatively low risk while more complex systems are required when highly toxic materials are used. A systematic way to identify the approach to be adopted for containing the hazards in any given application is described in the next chapter.

Development of a Containment Strategy

6

Purpose

The purposes of this chapter are twofold:

- to demonstrate how the hazard generated by a specific process operation is built up of many interrelated factors and not just the hazardous properties of the materials (chemicals) being used;
- to identify a range of Containment Strategies and how they can be used to meet the various types of exposure hazard.

Contents

Introduction

In order to select appropriate containment equipment to control a specific exposure hazard, it is helpful to be able to classify the performance of various types of equipment which, when combined with appropriate operation, maintenance and testing regimes, can provide an effective containment strategy for handling hazardous materials. It is important for the reader to understand that it is only through careful design, operation, maintenance and testing of containment equipment that the desired levels of performance will be achieved. Compromise of any one of these factors will result in under-performance of the containment system.

Quantification of the hazard

As explained in Chapter 3, it is important to identify the hazards associated with chemicals and assess the risks that they pose. The approach summarized in Figure 6.1 for quantifying hazards applies to both solids and liquids (including the vapours they produce). The key parameters considered are:

- hazard severity (e.g., toxicity);
- scale and duration of operation;
- physical properties (volatility or dustiness).

From these basic items of information, a **Containment Strategy** can be chosen. This chapter describes the various Containment Strategies available. Chapter 7 links specific items of equipment to these Containment Strategies.

The approach described in this chapter is based on recommendations by the Health and Safety Commission's (HSC) Advisory Committee on Toxic Substances, which also form the basis of HSG193.

Hazard severity

Many substances have been identified as hazardous and the severity of the hazards they present is identified by one or both of:

- an **exposure limit**, which may be one of the following, as defined in Chapter 2:
  — a Maximum Exposure Limit (MEL); or
  — an Occupational Exposure Standard (OES);
  — a Permissible Exposure Level (PEL);
- a Risk Phrase (R-phrase), which defines in standard phraseology the hazard(s) presented by that substance.

The Health and Safety Executive (HSE) publish lists of MELs and OELs annually as EH40. PELs are defined in 29 CFR 1910.1000. The standard R-phrases available are listed in Appendix 3 to this guide.

A substance may have a combination R-phrase made up of more than one basic R- phrase as a result of several toxicological properties. In such cases the substance is allocated for the purposes of selecting a Containment Strategy to the most stringent hazard group identified by any of the constituent basic phrases concerned. For some substances EH40 also identifies mutagenic or carcinogenic categories and both 29 CFR 1910.1000 and EH40 give a 'skin designation' (SK) to substances with a tendency to cause harm if absorbed through the skin (Sk).

Table 6.1 (page 64) shows how the exposure limit can be used to allocate a given substance to a **hazard group**. If no exposure limit is available, the R-phrase may be used instead; typical basic R-phrases for each hazard group are also shown in Table 6.1.

Where both exposure limits and R-phrases are available, they will generally identify the same hazard group. Allocation according to the most stringent exposure limit available, depending on duration of exposure, is the preferred method. However, it is up to those

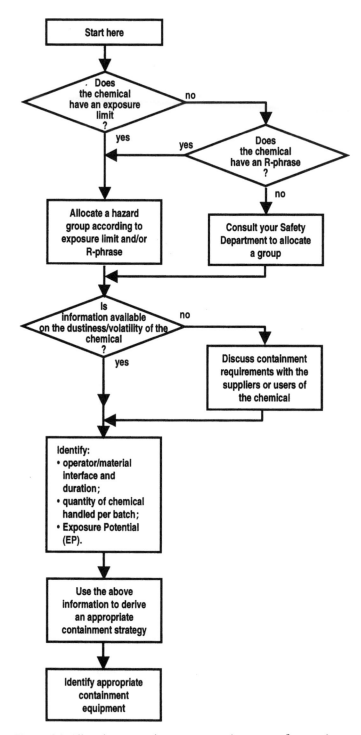

Figure 6.1 Allocating a containment strategy (summary of process)

Table 6.1 Allocation of hazard groups using exposure limit or R-phrases

Hazard group	Exposure limit	Typical basic R-phrases
A	1000–10000 $\mu g\,m^{-3}$ dust 50–500 ppm vapour	R36, R38
B	100–1000 $\mu g\,m^{-3}$ dust 5–50 ppm vapour	R20, R21, R22 (except in combination with R48)
C	10–100 $\mu g\,m^{-3}$ dust 0.5–5 ppm vapour	R23, R24, R25 (except in combination with R48) R34, R35, R37, R41, R43 R48 with any one or more of R20, R21, R22
D	1.0–10 $\mu g\,m^{-3}$ dust 0.05–0.5 ppm vapour	R26, R27, R28 Carc cat 3 R40, Muta Cat 3 R40 R48 with any one or more of R23, R24, R25 R60, R61, R62, R63
E	0.01–1.0 $\mu g\,m^{-3}$ dust 0.005–0.05 ppm vapour	R42, R45, R46, R49
F	$< 0.01\ \mu g\,m^{-3}$ dust; < 0.005 ppm vapour	No R-phrases assigned

selecting the containment equipment to satisfy themselves that the hazard group and indeed containment level allocated is the appropriate one for the substance concerned.

In addition to the groups listed in Table 6.1, HSG193 defines a *Group S* for those substances that can cause harm through contact with the skin or eyes. Such substances should be considered as belonging to Group S as well as the group allocated as described above. A substance should be allocated to Group S if either:

- it has a 'skin designation' ('SK') in 29 CFR 1910.1000 or EH40;
- its R-phrase includes any of R21, R22, R24, R27, R34, R35, R36, R38, R41, R43.

A full list of approved basic R-phrases is given in Appendix 3.

The method for allocating a substance to a hazard group according to HSG193 is based on the following principles:

- The hazard groups range from Group A (least hazardous) to Group E (most hazardous). Table 6.1 includes an additional Group F (which is a subset of Group E in HSG193), since materials of this toxicity are handled, albeit not commonly, in the pharmaceutical industry.
- The allocation to a particular hazard group of each R-phrase that denotes a toxicological property takes into account three factors:
  - (a) whether or not the toxicological end point will have an identifiable dose threshold (e.g., liver damage will, but genotoxicity will not);
  - (b) the level of seriousness of the resultant health effect (e.g., effects on foetal development are more serious than mild eye irritation);
  - (c) the exposure levels at which toxic effects occur compared with other substances (i.e., the relative potency of the substance in relation to a particular toxicological endpoint);

- With respect to (c) above, for certain toxicological endpoints, the current EU classification system identifies *dose criteria*, which take potency into account in assigning R-phrases. Dose criteria are applied in the classification and labelling of substances in relation to:
  - *acute toxicity*, with three categories of potency, defined as 'harmful', 'toxic' and 'very toxic';
  - *chronic toxicity* on repeated exposure, with two categories of potency, 'harmful' and 'toxic';
- Therefore, for these endpoints, potency has been partially quantified and this has been taken into account in the method described here.

However, this classification alone may not be enough. The General COSHH ACOP, L5 states that:

'The absence of a substance from the lists of MELs and OESs does not indicate that it is safe. In these cases, exposure should be controlled to a level to which nearly all the working population could be exposed, day after day at work, without adverse effects on health.'

It must also be remembered that this method is not intended to serve as a process for setting health-based occupational exposure limits but rather to assist in the design and selection of equipment and process systems.

Scale of operation

The scale of operation, i.e., the quantity of material transferred in a discrete operation, potentially affects the quantity of released material and hence the risk of exposure. As indicated in Chapter 4, production objectives can require transfers varying in quantity from a few kilograms to more than 1 tonne. An approximate scale of operation, for risk assessment purposes, is shown in Table 6.2.

Table 6.2 Scales of operation

Operation	Scale
Lab/pilot transfers — gm–kg (solids) or 1–1000 millilitres (liquids)	Small
Transfers in range 10–100 kg (solids) or 1–1000 litres (liquids)	Medium
Large-scale production transfers — over 100 kg (solids) or 1000 litres (liquids)	Large

For example, many transfer operations involving quantities below 100 kg may be required to fill a 1000 kg Intermediate Bulk Container (IBC). Alternatively, the same quantities may be moved by means of a single large- scale IBC transfer. The latter operation may have the potential to release more material than the series of small-scale transfers; however, it may be easier to achieve engineered containment for one large transfer than for a series of smaller ones.

Physical form of the material

The Containment Strategy to be adopted for a substance depends not only on its hazard group but also on its *Exposure Potential* (EP). This in turn depends on the dustiness (in the case of a solid) or the volatility (in the case of a liquid) as well as on the scale of use and duration of exposure. Clearly, dusty and volatile materials are more likely to spread out to the environment than very granular solids and liquids of low vapour pressure. Hence a qualitative scale of the potential exposure risk due to the physical properties of the material may be applied. The allocation of Containment Strategies will therefore be considered separately for solids and liquids, in the next two sections.

Allocating a Containment Strategy – solids

Table 6.3 gives an empirical method of categorizing the dustiness of a powder.

Table 6.3 Dustiness potential

Dustiness	Characteristics
Low	Pellet like, non-friable solids Little dust seen during use *Examples: PVC pellets, waxed flakes, pills*
Medium	Crystalline, granular solids Dust is seen during use but settles out quickly Dust is seen on surfaces after use *Example: Soap powder*
High	Fine, light powders When used, they form dust clouds that can be seen and that remain airborne for several minutes *Examples: cement, carbon black, chalk dust*

Combining the scale and duration of an operation and the dustiness potential of a powder it uses gives the EP, as shown in Table 6.4.

Table 6.4 Exposure potential — solids

Dustiness potential Quantity handled	Low	Medium	High
Small (gm)	EP1 / EP1	EP1 / EP2	EP2 / EP3
Medium (kg)	EP1 / EP2	EP2 / EP3	EP3 / EP4
Large (tonnes)	EP2 / EP3	EP3 / EP4	EP3 / EP4
Task (transfer) duration:	Short / Long		

The classification of task duration (i.e., the period of open transfer or potential for exposure) as 'short' or 'long' is a matter for judgement. Generally an operation lasting less than 30 minutes can be considered 'short'.

The exposure potential thus derived, when associated with the hazard group of the material, can be used to allocate a Containment Strategy appropriate to the risk of the operation, as shown in Table 6.5. Descriptions of the strategies are given in Table 6.8 (page 69) with further information in Table 6.9 (pages 70–71).

Table 6.5 Allocation of Containment Strategy — solids

| Hazard group | Exposure potential | | | |
	EP1	EP2	EP3	EP4
A	Strategy 1	Strategy 1	Strategy 1	Strategy 2
B	Strategy 1	Strategy 2	Strategy 2	Strategy 3
C	Strategy 2	Strategy 3	Strategy 3	Strategy 4
D	Strategy 3	Strategy 3	Strategy 4	Strategy 4
E	Strategy 4	Strategy 4	Strategy 4	Strategy 4
F	Strategy 5	Strategy 5	Strategy 5	Strategy 5

Allocating a Containment Strategy – liquids

The classification system for volatility recommended by the HSC's Advisory Committee on Toxic Substances is shown in Figure 6.2 (page 68). As in the case of solids, described in the previous section, the exposure potential of a hazardous liquid is affected by its volatility and the scale of the operation that uses it. The EP is allocated as shown in Table 6.6 (page 68), based on 'long' or 'short' exposure durations as for Table 6.4.

The exposure potential thus derived, when associated with the hazard group of the material, can be used to allocate a Containment Strategy as shown in Table 6.7 (page 68). Descriptions of the strategies are given in Table 6.8 (page 69) with further information in Table 6.9 (pages 70–71).

Definition of Containment Strategies

A *strategy* in this context is a technique, recognized by the HSE and the occupational hygiene profession, for managing hazards in the workplace. It embraces both the management issues and the engineering devices that are necessary for the safe handling of hazardous chemicals. The Containment Strategy therefore has three essential components:

- equipment design and operating procedures;
- training, in:
  — access;
  — operation;
  — emergencies;
- maintenance.

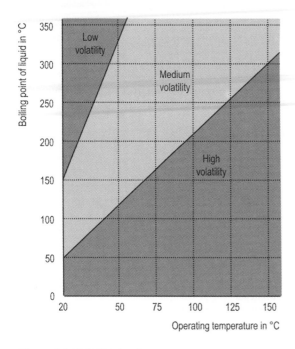

Figure 6.2 Volatility levels

Table 6.6 Exposure potential — liquids

Volatility / Quantity handled	Low	Medium	High
Small (ml)	EP1 / EP1	EP1 / EP2	EP2 / EP3
Medium (litres)	EP1 / EP2	EP2 / EP3	EP3 / EP3-4
Large (m³)	EP2 / EP3	EP3 / EP4	EP3 / EP4

Table 6.7 Allocation of Containment Strategy — liquids

Hazard group	Exposure potential			
	EP1	EP2	EP3	EP4
A	Strategy 1	Strategy 1	Strategy 1	Strategy 2
B	Strategy 1	Strategy 2	Strategy 2	Strategy 3
C	Strategy 2	Strategy 3	Strategy 3	Strategy 4
D	Strategy 3	Strategy 3	Strategy 4	Strategy 4
E	Strategy 3	Strategy 4	Strategy 4	Strategy 4
F	Strategy 5	Strategy 5	Strategy 5	Strategy 5

Table 6.8 Containment Strategy equipment

Strategy 1: Controlled general ventilation 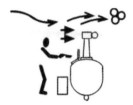	No special engineering requirements; adequate control is achieved by general ventilation of the process area. (This strategy is not covered further in this guide)
Strategy 2: Local exhaust ventilation 	A Local Exhaust Ventilation (LEV) system is used to contain the contaminants within a defined area and draw airborne contaminants away from the operators' breathing zone. This can involve either: • a good point exhaust ventilation; or • a unidirectional air-flow booth. This can achieve significant reductions in operators' exposures to the concentrations of airborne dusts and vapours generated during open transfer operations of hazardous materials.
Strategy 3: Open handling within isolator or **High-integrity closed coupling without external containment**	Open transfer or handling of hazardous materials takes place within an isolator. Typically this might involve surrounding the transfer operation with a fixed or flexible air-tight barrier. Containers of process material may be placed in or removed from the isolator only in a way that does not compromise the integrity of the containment it provides. The operator uses a glove-port to effect the transfer of material to or from the open container and to clean empty containers. This Containment Strategy can also cover transfers effected by means of a high-integrity coupling between closed containers without an external isolator.
Strategy 4: Closed handling within isolator	Closed transfer or handling of the hazardous material takes place within an isolator. This is similar to the preceding strategy except that open transfer is not permitted even within the enclosure. The operator, again using a glove-port or similar device, attaches the closed container directly to the access port for the process to form a closed connection and then opens the valve to effect the transfer of material.
Strategy 5: Robotic handling, contained system 	This strategy is adopted for materials so hazardous that even with a closed transfer system the use of a glove-port represents an unacceptable risk because of the possibility that the gloves could rupture. The transfer therefore has to be effected by a fully automated enclosed process. The strategy requires highly specialized training and should be prepared and implemented only after consultations with experienced health and safety professionals and the HSE.

Table 6.9 Containment Strategy considerations — operational access

Strategy 2	Strategy 3	Strategy 4	Strategy 5
Only authorized personnel should be allowed into the working area.	Entry to the working area should be controlled.	Entry to the working area must be controlled.	No entry to the process area will be permitted in normal operation.
	Work areas and equipment should be clearly signposted.	Work areas and equipment must be clearly signposted.	All operator input to the process will be via remote control.
Training must be provided on decontamination procedures to be followed prior to leaving the work area.	Special training on emergency evacuation and rescue must be provided.	Only operators trained in emergency evacuation procedures will be allowed access to the area.	No entry is permitted, even in an emergency: remote-controlled or automated emergency systems are needed.
LEV must be applied at source to capture contaminants. This must be carefully designed and sited so that its performance is not compromised by external features such as draughts from doors or the building's general ventilation system.	Plant and equipment must be totally enclosed to a standard normally encountered in an industrial environment.	The process operation must be totally contained by equipment specially designed for the purpose.	The process operation must be totally contained within multiple layers of containment designed specifically for this purpose by specialist designers. Such a system may, for example, comprise a totally sealed process with fully welded pipe connections operating within a sealed plant envelope.
Whilst the systems need not always be closed, the efficiency of the LEV will be improved by partially enclosing the source (e.g., in a fume cupboard or booth) and this degree of containment may be necessary for some applications.	Only limited breaching of containment, e.g., for taking samples, is permitted.	No breaching of containment in operation is permitted.	No direct operator interface or breach of containment is permitted.

(continued)

Table 6.9 (*Continued*)

Strategy 2	Strategy 3	Strategy 4	Strategy 5
Relative location of operations and LEV should prevent escape of contaminants into the general working area.	Enclosures should be maintained under negative pressure to prevent leakage.	Enclosures should be maintained under negative pressure to prevent leakage.	Enclosures must be fitted with secondary envelope, both maintained under negative pressure to prevent leakage.
Exhausted air may be recirculated only if first cleaned by a high-capacity filter backed up by a safe-change High-efficiency Particulate Arrestor (HEPA).	Contaminated air from the extraction system should be passed through a suitable safe-change HEPA before being exhausted outside the building.	Contaminated air from the extraction system must be passed through a suitable safe-change HEPA before being exhausted outside the building.	Contaminated air from the extraction system must be passed through at least a double safe-change HEPA before being exhausted outside the building.
A regular preventive maintenance programme should be implemented for air extraction systems.	Regular certification and testing of the filtration system will be required.	Regular certification and testing of the filtration system will be required.	The filtration system must be backed up by a second system. Regular certification and testing of both systems is required.
Operator manipulates compounds directly. PPE may be required.	Operator manipulates compounds via glove-box interface.	Operator may prepare containers for transfer direct from container to vessel.	Containers for transfer must be prepared by robot control in an enclosed process.

It is important that designers and purchasers of containment systems should understand these three aspects.

Table 6.8 provides an overview of the intent of five **Containment Strategies** and the equipment required to implement them. The strategies considered further in this guide are 2, 3, 4 and 5.

Containment Strategies 1 and 2 above equate to Control Approaches 1 and 2 in HSG193. Containment Strategies 3 to 5 between them equate to Control Approaches 3 and 4 but without a direct one-to-one mapping.

The following tables compare in more detail the management aspects of each of the strategies as follows:

Table 6.10 Containment Strategy considerations — personal protective equipment (PPE)

Strategy 2	Strategy 3	Strategy 4	Strategy 5
RPE is not normally required, but there may be occasional short-term activities where its use is needed.	RPE is not normally required. There may be specific short-term activities where high-efficiency RPE will be needed.	RPE is not normally required. Under emergency conditions high- efficiency RPE will be needed.	Pressurized air-suits may be necessary in areas adjoining the process area during emergencies.
PPE such as protective overalls, gloves and goggles will be required for maintenance procedures and may also be required for normal handling of materials.	Special PPE will be required for equipment break-down and maintenance.	Impervious overalls, gloves and eye protection should be worn for equipment break-down and maintenance. Full PPE (including suitable RPE) is needed if the enclosure is opened when contaminated.	PPE other than pressurized air-suits is not recommended.

- access during normal operations (see Table 6.9);
- Personal Protective Equipment (PPE), including Respiratory Protective Equipment (RPE) (see Table 6.10);
- maintenance and cleaning (see Table 6.11, page 74);
- training (see Table 6.12, page 75).

Verifying the selected strategy

Previous sections of this chapter have considered how to use the combination of the hazard group, scale of operation and dustiness/volatility to select a Containment Strategy. Chapter 7 describes how to select containment devices from a list of generic types to meet the requirements of each of these Containment Strategies. Chapter 10 describes how the performance of the containment devices should be validated once they have been installed.

Before any decision is made to purchase a particular device or item of containment equipment, those responsible are urged to verify the choice of Containment Strategy by seeking advice, as listed in Table 6.13 (page 76), on each stage of the process followed. Further suggestions as to whom to consult regarding each aspect of the selection process are given in Table 6.14 (page 76). It is also essential to base the choice on a thorough risk assessment that considers not only the parameters already discussed but also the overriding constraints (see next section) and any operational factors that may be relevant.

The purchasers must remember to:

- ask the suppliers how they handle the materials in question;
- consider the choice of container and whether the size/type can be changed;
- consider the use of alternative materials, e.g., replace hazardous substances with less hazardous alternatives or use other physical forms of the same materials;
- challenge all operations, so that unnecessary transfers can be identified and eliminated.

Overriding constraints

Earlier, it was mentioned that not all hazards of materials are reflected in exposure limits or R-phrases. Certain compounds (e.g., explosives, radioactive materials, extreme biohazards, sensitizers, carcinogens, mutagens and any substance in Group S as defined on page 64) require special engineering and operational controls. Whilst in many cases these can be adequately controlled by Containment Strategy 4 or 5, they may also require specialist containment techniques beyond the scope of this guide.

Conclusion

This chapter has outlined the aims of five Containment Strategies and described the method recommended for identifying the strategy appropriate to a given application. Various exposure parameters are suggested for use in applying this method: where the produce conflicting results or results close to the boundary criteria for two containment strategies, it is recommended that the more stringent strategy should be chosen. Selection of the strategy must be made in the context of a full risk assement and the responsibility rests with those making the selection to satisfy themselves that the limits chosen are appropriate. The next chapter describes how these aims can be achieved in practice.

Table 6.11 Containment Strategy considerations — maintenance and cleaning

Strategy 2	Strategy 3	Strategy 4	Strategy 5
Surface finishes must be easy to clean and non- porous.	Surface finishes must be crevice-free and ground smooth to facilitate cleaning.	The highest standard of surface finish is required. This must be compatible with automated cleaning such as Cleaning-in-Place (CIP) systems.	Routine cleaning of the fully sealed process plant will not usually be required. Automated decontamination of the process will be necessary prior to any entry into the process area. Full PPE may be required even after decontamination.
A regular maintenance and cleaning schedule for equipment and surfaces should be implemented.	A regular maintenance and cleaning schedule for equipment and surfaces should be implemented.	A regular maintenance and cleaning schedule for equipment and surfaces must be implemented.	Regular inspection and maintenance may be a legal requirement; specialist assistance must be sought.
A good standard of housekeeping is expected. Cleaning should be by vacuum or wet mopping. Normal maintenance procedures should be followed.	Equipment design should facilitate easy maintenance and cleaning.	Equipment design must include automated maintenance and cleaning facilities.	Thorough decontamination of the plant must be carried out and verified before maintenance staff enter the plant enclosure. Specialist advice must be obtained.
Dry brush sweeping and compressed-air cleaning should be avoided.	Special procedures, such as purging, or cleaning procedures, such as CIP, must be implemented before systems are opened.	Special procedures, such as purging, or cleaning procedures, such as CIP, must be implemented before systems are opened.	Permit-to-work systems with regular medical surveillance of operators and maintenance staff may be legal requirements. Specialist advice must be obtained.
PPE (including suitable RPE) is likely to be required when equipment is opened for maintenance.	Permit-to-work systems should be considered for maintenance activities. PPE (including suitable RPE) is likely to be required when equipment is opened for maintenance.	Permit-to-work systems must be considered for maintenance activities. PPE (including suitable RPE) is likely to be required when equipment is opened for maintenance.	Maintenance may take place only after decontamination and when authorized by permit to work. PPE (including suitable RPE) will be required when equipment is opened for maintenance.

Table 6.12 Containment Strategy considerations — training

Strategy 2	Strategy 3	Strategy 4	Strategy 5
In addition to basic induction training, specific training is required on: • the hazardous nature of the substances handled; • the operation of the process. Particular attention should be given to how to: • detect and respond to a failure in the ventilation system; • carry out open transfers so as to minimize dust release in the breathing zone.	Specific on-the-job training is required. This should include an understanding of: • the plant; • the hazardous nature of the substances handled; • the operation of the process and equipment; • maintenance procedures; • use of PPE; • procedures to detect and deal with loss of containment.	Emergency response training is required. This should include an understanding of: • the plant; • the hazardous nature of the substances handled; • the operation of the process and equipment; • maintenance procedures; • use of PPE; • procedures to detect and deal with loss of containment in an emergency.	Operator training and emergency response procedures must be set up and reviewed on a regular basis, with retraining provided as soon as the need is identified and regular refresher training provided routinely. Full liaison with the HSE is recommended at the planning stage and in the light of significant findings of a review.
Periodic retraining/refresher training will be required.	Periodic retraining/refresher training will be required.	Periodic retraining/refresher training will be required.	

Table 6.13 Verification of Containment Strategy

Selection criterion	Verification method
Hazard group and material characteristics	Seek the advice of the manufacturer/supplier.
Specific task exposure monitoring	Seek the co-operation of the safety department.
R-phrase evaluation	Seek the co-operation of the safety department.
Scale of operation	Discuss all aspects with production/maintenance to ensure that each potentially hazardous task is covered by the containment strategy.

Table 6.14 Input guide for strategy selection

Input elements	Chemical supplier	Occupational hygienist	Health and safety development	Operations	Maintenance group	Quality assurance validation	Equipment/ system designer
Hazard group	✓	✓	✓				
Scale of operation				✓			
Exposure potential		✓	✓	✓	✓		
Frequency and task duration				✓	✓		✓
Operability of device			✓	✓	✓	✓	✓
Cost of device	✓			✓	✓		✓

Containment equipment types

7

Purpose

The purpose of this chapter is to enable the reader to select appropriate equipment and operating systems to implement the Containment Strategy identified for the problem in hand as described in Chapter 6. It includes a series of containment device evaluation sheets detailing the main characteristics and applications of the various control devices available.

Contents

General principles

As Chapter 6 demonstrated, the choice of engineering control device to contain the hazardous substances in a specific process or operation is based on:

- hazard band, derived from exposure limit or R-phrase;
- scale of operation;
- dustiness or volatility of the material;
- duration of operator interface.

Chapter 6 also defined a method for categorizing such devices under five **Containment Strategies**, although it stated that this guide does not consider Containment Strategy 1. This chapter discusses how the remaining four strategies can be implemented in practice. Table 7.1 lists the typical characteristics of the generic engineering control devices appropriate to each containment.

Table 7.1 Characteristics of control devices for each Containment Strategy

	Strategy 2: Local exhaust ventilation	Strategy 3: Open handling within isolator	Strategy 4: Closed handling within isolator	Strategy 5: Robotic handling, contained system
Solids	Directional airflow booth.	Glove-box isolator. Operator physically separated from materials being handled.	Direct powder or liquid transfer connections within barrier isolator.	Robotics/remote operation.
Liquids	Exhaust hood. Drum booth with PPE. Shielded dip pipe.	Pipe couplings or hoses. Connections made or broken inside closed isolator.	Pipe couplings or hoses. Contained breakages.	Isolated pipe line system, pipes within pipes, etc.

These designs differ in costs, performance benefits and user interface requirements. Table 7.2 lists some of the performance characteristics and benefits of the four strategies considered.

Within these parameters, all these engineering control devices differ in applicability with regard to:

- the scale of operation;
- material exposure potential;
- suitability for high-toxicity materials.

Containment Strategy 2 (air-flow containment) devices do not rely on specific coupling devices but achieve adequate control for low-hazard materials largely as a result of removal of the airborne contaminants in a unidirectional air-stream, provided that the operator follows suitable procedures. Most contained operations (Containment Strategy 3) rely on the ability to transfer materials, components and parts into and out of an isolator or between vessels contained within it without releasing significant quantities of airborne hazardous materials into the environment. Containment Strategy 4 systems require closed transfer or coupling devices to be used whose connections are made and broken within an isolator. In totally

Table 7.2 Key features of Containment Strategies

	Strategy 2	Strategy 3	Strategy 4	Strategy 5
Performance	Effective for manual powder transfer operations. Needs high velocity local dust capture for best results.	High levels of operator protection. Weak links: • glove porosity; • entry/exit ports; • difficult ergonomics;	Higher levels of operator protection. May require automation. May not be easily adaptable to meet variations in the process.	Maximum levels of operator protection in use, since it relies on remote control by the operator using it. Decontamination necessary prior to entry by maintenance crew.
	Personal Protective Equipment (PPE) may be needed in addition when handling certain hazardous materials.	No PPE needed for normal operations.	No PPE needed for normal operations.	No PPE needed for normal operations.
Benefits	Flexible and operator-friendly. Can form an engineered workstation.	Operator safety. Contained work-space. Suitable for clean-down using Cleaning-in-Place (CIP) techniques.	High levels of containment assured. Operator interface minimized. Suitable for clean-down using CIP techniques.	Layering of containment systems permits attainment of high protection factors. Suitable for clean-down using CIP techniques.

enclosed systems (Containment Strategy 5) all operations are carried out by robotics; the operator has no contact at all with the hazardous material but undertakes all required actions by remote control, normally from outside the area or room in which the transfer operation, inside its own isolator, takes place. The key features of devices suitable for these Containment Strategies are described below.

Containment Strategy 2 – Local exhaust ventilation

Containment Strategy 2 depends on a directional air flow to control the quantities of dust and vapours that could escape into the operator's breathing zone. The simplest form of device suitable for this strategy comprises an air

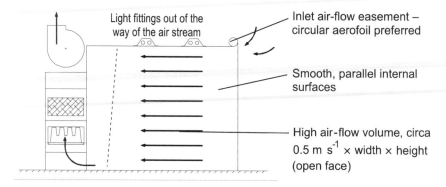

Light fittings out of the way of the air stream

Inlet air-flow easement – circular aerofoil preferred

Smooth, parallel internal surfaces

High air-flow volume, circa $0.5\ \mathrm{m\ s^{-1}} \times$ width $\times$ height (open face)

Figure 7.1 Horizontal laminar flow booth principle (exhaust only)

exhaust hood or some other form of Local Exhaust Ventilation (LEV), which draws air away from the operator. Protection is improved considerably if the source generating the dust or vapour is partially enclosed, as this will minimize the adverse effects of draughts and air movement and avoid spreading contamination within the plant. The air should flow in a uniform direction away from the operator so as to minimize eddies and turbulence. Collected pollutants may require dry filtration or scrubbing, depending on their nature and explosion potential.

Figure 7.1 and Figure 7.2 illustrate two applications of this principle. In the *horizontal laminar flow booth* (see Figure 7.1) air is exhausted from one end of the booth through a perforated distribution plate, which ensures that an even flow of air is drawn into the booth from the open end.

In the *down-flow booth* (see Figure 7.2) air is supplied into the booth via the ceiling and out through a vent low down at the back. In many types of down-flow booth the exhausted air is cleaned and recirculated: up to 90% of it back into the booth and the remainder out into the open workplace. The cleaning system should comprise a pre-filter with high dust-holding capacity supplemented by a High-efficiency Particulate Arrestor (HEPA). The fans used

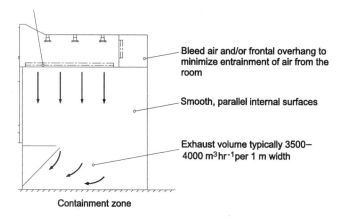

Bleed air and/or frontal overhang to minimize entrainment of air from the room

Smooth, parallel internal surfaces

Exhaust volume typically 3500–4000 $\mathrm{m^3\,hr^{-1}}$per 1 m width

Containment zone

Figure 7.2 Down-flow booth principle (supply and exhaust air flow)

should be able to move large volumes of air without generating excessive noise levels. These considerations, together with the need to fit cooling coils to dissipate the heat generated by the fans, add to the cost of this type of booth. However, this increase must be offset against the cost of replacing large volumes of air drawn only once through other types of booth and then discharged outside the plant. Also, where air is cleaned and recirculated in this way, the air in the booth is often cleaner than the environment outside it.

For some cases where acceptability of Containment Strategy 2 is borderline but Containment Strategy 3 may not be reasonably practicable, it may be acceptable to use Containment Strategy 2 but require employees to wear masks, goggles, gloves, lab coats or other appropriate forms of PPE whilst in the booth. Indeed, even where Containment Strategy 2 theoretically provides adequate containment by itself, the use of such PPE may be advisable as a precaution against the disturbance of the air flow by the operator's position or movements. However, if a full air-suit with respirator would be routinely required to provide adequate protection during normal operations, then Containment Strategy 2 is not appropriate and some form of enclosure in accordance with Containment Strategy 3 or 4 should be considered. Chapter 9 gives further guidance on the use of PPE.

Containment Strategies 3 and 4 – Open/closed handling within isolator

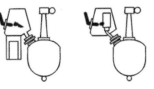

Both Containment Strategies 3 and 4 require hazardous materials to be handled only within isolators. The design of such systems must therefore consider not only the isolator itself but how containers of hazardous materials may be transferred into or out of it without releasing contaminants into the environment.

In the case of Containment Strategy 3, the materials may be handled openly within the isolator. For Containment Strategy 4, however, material transfers may be effected only through closed systems whose connections are made and broken within the confines of an isolator. Depending on their design, the coupling devices used to make these connections may also provide adequate protection for Containment Strategy 3 without the added protection of an isolator.

These essential ingredients of a containment system for strategies 3 and 4 — isolators, transfer devices and coupling devices — are discussed in turn below.

Isolators

With Containment Strategy 3 equipment, the operator is separated from the hazardous solids or liquids by a fixed or flexible barrier. Fixed barriers normally comprise a solid steel housing with glazed viewing panels; flexible barriers typically comprise welded transparent plastic enclosures sealed around the process. Both types generally incorporate a further type of flexible barrier: a plastic membrane in the form of gloves or a half-suit built into the container, as shown in Figure 7.3 (page 82). Within such a closed isolator, materials (solids or liquids) can be handled openly: for example, they may be moved between vessels by means of scooping or pouring.

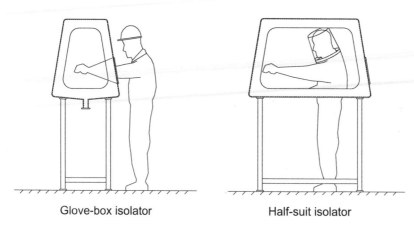

Glove-box isolator   Half-suit isolator

Figure 7.3 Typical CS3 isolator designs

The isolator chamber must be ventilated so that residues of hazardous substances can be removed and airborne concentrations reduced to acceptable levels before the door(s) of the isolator are opened to enable the operator to remove an empty container or replace it with a full one. Advantages of this arrangement are that air-flow volumes are minimized and nitrogen purge systems can easily be used to minimize explosion risk, as described in Chapter 8. Interlocks must be used to ensure that the supply routes for the hazardous substances are closed before the isolator door(s) can be opened and remain closed until the doors are closed again.

As the operator manipulates the hazardous materials via glove-ports or half- suit barriers, it must be verified prior to use that these are made of materials that will withstand the substances concerned. Such considerations are described later in this chapter, from page 89.

Ergonomic acceptance of the design selected is the main challenge, whether small weights of materials are manually transferred or heavier drums manipulated.

Design features for both types of isolator should include the following:

- fully welded shell with coved corners, or large plastic envelope;
- welded-in pipe connections;
- built-in provisions for draining;
- safe entry/exit port for materials;
- glazing flush with walls on the inside to prevent accumulation of contaminant;
- good ergonomics — consider mock-up for evaluation or approval of the design by users;
- safe-change filtration elements.

Transfer devices

The simplest device for containing the operation of loading the contents of a drum into a process vessel is a glove-box chamber with its own ventilation system and a side door to admit the drum as shown in Figure 7.4(a). Once the drum is inside the chamber, the door is closed and the drum is emptied using open transfer, with most of the dust generated in the atmosphere

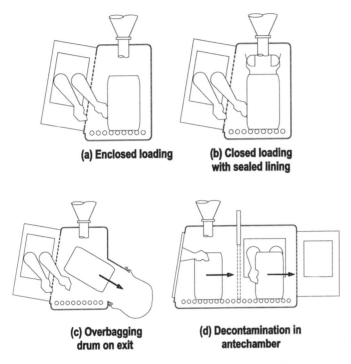

(a) Enclosed loading

(b) Closed loading with sealed lining

(c) Overbagging drum on exit

(d) Decontamination in antechamber

Figure 7.4 Containment techniques for loading and unloading drums

within the chamber removed by the ventilation system. When the operation is complete and airborne contamination reduced to an acceptable level, the door is opened and the drum removed. A single door thus provides partial protection for the operator but may become contaminated on the inside, as may the outside surface of the drum.

One way to minimize the spread of contamination outside the glove-box chamber when the door is opened would be to use a drum fitted with an internal plastic lining and seal this lining to the duct that penetrates the chamber for loading material into the process as shown in Figure 7.4(b) and described further on page 99. This would reduce the amount of airborne dust generated in the isolator and keep its door and the surfaces of the drum free from contamination. Such devices can achieve containment levels approaching those of the well engineered closed-coupling valves described on page 85, though they are specifically designed for transferring materials into or out of drums with liners.

Another commonly used method is to enclose the drum in a sealed bag before it leaves the chamber as shown in Figure 7.4(c). Devices available commercially dispense polythene tubes that are wide enough to enclose the drum and can be sealed around it and cut to fit. A simplified version of this technique is described overleaf.

As an alternative to either of the foregoing arrangements, or even in addition to it, the transfer chamber could be supplemented with an antechamber in which the drum could be cleaned before being removed as shown in Figure 7.4(d). The outside door of the antechamber could be opened with minimal contamination on its inner surface, since all transfer would take place with the inner door, leading to the transfer chamber, closed. However, this method relies on the use of correct procedures and failure on the part of operators or their supervisors to ensure that these are followed at all times may jeopardize safety.

Reasonably contained filling of drums without liners can be achieved in double-chamber isolators by raising the drum so that its lip is sealed against the filling orifice of the transfer chamber. Before the full drum is lowered, the operator fits a seal to this filling orifice, using a glove-port to work inside the chamber.

Overbagging

The use of a bag to cover the removal of items from an isolator, be they waste or contaminated containers as shown Figure 7.4(c), is an important technique with a variety of applications. The technique in one of its simplest forms, to remove a contaminated object from an isolator, is illustrated in Figure 7.5.

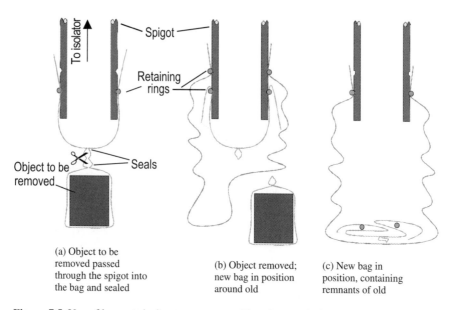

(a) Object to be removed passed through the spigot into the bag and sealed

(b) Object removed; new bag in position around old

(c) New bag in position, containing remnants of old

Figure 7.5 Use of bag-out device to remove an object from an isolator

The object to be removed is passed into a bag through a spigot with two grooves around its circumference. A rubber retaining ring holds the bag in place by engaging with one of the grooves — normally the one nearer the end of the spigot, as shown in Figure 7.5(a) — to form a seal sufficient to satisfy Containment Strategy 3.

When the object is ready to be removed, the bag is tied around it with two seals close together and cut between the ties. An improvement on this technique is to use heat to seal the bag in three places and to cut across the middle seal. The sealed bag containing the object can then be removed. A new bag is placed over the remnant of the old one around the spigot and secured by another ring, engaging with the groove nearer the isolator (b). Working through the

84

new bag, the operator disengages the old one from the spigot, allowing its loose end and retaining ring to fall into the new bag (c). Meanwhile, the ring holding the new bag is moved to engage with the groove nearer the end of the spigot and the new bag is then ready to receive further items for removal.

In some versions of this application the bags are formed from lengths of tough but flexible polythene tubing stored along the spigot, as shown on page 110. A similar technique can be used to remove contaminated filters and insert clean replacements, as described on page 111. However, the procedures involved can be quite complex and it is essential to follow the manufacturer's instructions.

Coupling devices

Transfer of materials requiring protection in accordance with Containment Strategy 4 must be effected by closed connections between the source and destination vessels. These connections must be made or broken under cover of a glove-box or half-suit isolator as described earlier. In some cases, the same devices can provide adequate protection to satisfy Containment Strategy 3 even if the connections are made and broken in the open workplace.

Rapid transfer ports

For a coupling system to qualify as an effective 'closed' transfer device, it must effect the transfer through a sealed and contained system without compromising the containment of the hazardous materials. The first example of this was the Double Porte de Transfer Entaché (DPTE®), patented by La Calhéne, Inc. in 1960. It was a milestone in the development of today's isolator. Since then, many similar devices have been developed, which are now known as **Rapid Transfer Ports** (RTPs) or **alpha-beta couplings**. Clearly, these devices need to be subjected to rigorous testing and maintenance protocols during the life of the isolator, so as to ensure that opportunities for breach of containment do not occur. RTPs allow the rapid transfer of relatively small quantities of materials and components (both preferably bagged) through a barrier wall, thereby ensuring the protection of operations and operators from exposure to the hazardous environment within the isolator.

The basic principle of the RTP relies on the integration of four basic components that provide a dust-tight or vapour-tight seal and effect a tight connection:

- the isolator door;
- the isolator door frame;
- the container valve;
- the container valve frame.

The principle of operation is illustrated in Figure 7.6 (page 86). In this example a hazardous material that is to be dispensed into smaller batches is supplied in a drum sealed with an RTP fitting. The weighed batches are to be placed in a smaller RTP-flanged container for transfer to the production process. To complete the charge sequence, the transfer container is offered up and docked onto the process charge isolator to enable the batches to be loaded safely into the process.

(a) Drum and transfer container positioned beside glove port

(b) Port doors open together; contents of drum weighed into batches

(c) Weighed batches placed in transfer container

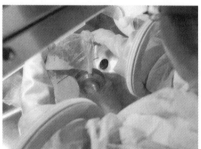

(d) Transfer container undocked after port doors have been closed

(e) Batches removed from transfer container being charged into the process

Figure 7.6 Use of RTP

Figure 7.6(a) shows the drum in place on the right of the isolator and the transfer container similarly docked on the left to receive the process batches. After the drum has been twisted or locked into place so as to engage the port, the two port doors, one fitted to the isolator, the other belonging to the drum, may be removed as a unit inside the isolator, allowing the material to be dispensed into small batches (b). Removing the doors together ensures that the outer surface of the drum port door is protected from contamination during the transfer. This is usually achieved by the use of extremely close fitting elastomer seals. The process is described further and illustrated on page 116.

The port door on the left of the isolator and that of the smaller container are similarly opened as a unit, allowing the weighed batches to be placed in the transfer container, (c). When all the batches required have been loaded into the container, the two port doors on the isolator and the transfer container will be replaced together, allowing the transfer container to undock, (d), leaving only clean, uncontaminated surfaces exposed.

The transfer container is then docked onto a similar port on an isolator protecting the opening of the process vessel and the two port doors opened as described above. The batches are removed as required from the transfer container and added to the process vessel, (e).

Developments in this technology have resulted in the introduction of interlocks such that the door cannot be opened unless a container has been properly connected and equally, the container cannot be removed if the door is not shut. It is also possible to obtain RTPs that can be coupled together without rotation.

Split butterfly valves

A limitation of the RTP is that it requires an isolator into which the coupled door of the container is opened. An RTP cannot connect a container directly, for example, to a reactor nozzle fitted with the other half of the RTP, because it is not possible to access the inside of the reactor to open it. As a solution to permit direct connections that will enable material to be transferred from small containers or Intermediate Bulk Container (IBC) bins directly into process vessels, *Split Butterfly Valves* (SBVs) in various forms have more recently found favour for use in bulk transfer operations.

A simple butterfly valve is a disc the diameter of the tube in which it is housed, capable of rotating about an axis across one of its diameters so that it either blocks the flow of fluid or powder or lies parallel to it, permitting the fluid or powder to flow past it. The split butterfly valve, as its name suggests, is effectively a single butterfly valve that is split into two parts, shown in Figure 7.7. One matching part (the *passive valve*) closes off and seals the bottom of the container holding the powder, whilst a matching half (the *active valve*) seals the inlet connection of the receiving vessel. Split butterfly valves must always be closed when the two parts are separated. Only when both valves are docked together to form a complete connection do interlocks release to permit the combined disc to rotate 90° to open and permit transfer.

(a) Passive valve   (b) Active valve

Figure 7.7 Split butterfly valve

Different valve designs use alternative docking methods. Several designs use bayonet pins either side of the valves, as shown in Figure 7.7, to centralize and position the two halves before bayonet connectors are twisted to clamp them together. Other designs use a clamp ring to achieve the same result. All designs require accurate positioning to effect a satisfactory 'dock'.

One benefit of the SBV design is that true direct connections between containers and vessels are possible. The RTP requires an intermediate chamber to move the port doors out of the transfer route. The difference is illustrated in Figure 7.8.

However, the containment afforded by SBV systems is not as effective as that of RTPs. SBVs should therefore not be considered on their own for materials with an exposure limit much below 10–$15\ \mu g\, m^{-3}$ dust in air; i.e., they are suitable for Band C and occasionally Band D materials as defined in Chapter 6. To achieve concentrations below this, secondary protection around the coupling zone is required.

Containment Strategy 5 – Robotic handling, contained system

Containment Strategy 5 is required when the substances involved are so hazardous that the operator must be isolated from any contact at all and even

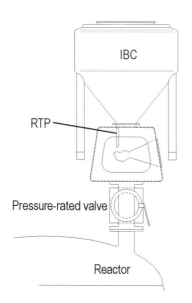

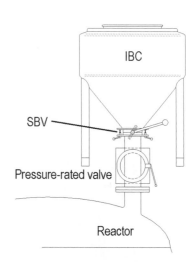

(a) Intermediate chamber required with RTP system

(b) Direct connection to vessel via SBV system

Figure 7.8 Connection using RTP and SBV arrangements

manipulation through a glove-box represents an unacceptable risk, as leakage could occur. Transfers must be effected through closed systems, whose connections must be made and broken only by remotely controlled robot systems.

Where the Containment Strategy selection is found to be at level 5, specialist suppliers must be used and extensive proving tests will be necessary to ensure that the required levels of operator protection are achieved. It will also be necessary to provide comprehensive operator training using a process simulator.

Materials used for constructing containment systems

Engineers designing a containment system should consider the chemicals that will be contained within the system. These chemicals could affect the materials used in construction in two ways:

- *degrading* the material, by corrosion, solvation or other changes that weaken its ability to keep chemicals out;
- *permeating* through it so that it no longer acts as a barrier.

Strong acids and oxidizing agents often cause degradation, whereas organic chemicals will permeate through many of the polymeric materials used in construction. While metals will act as an effective barrier, natural and man- made polymers such as PVC, nitrile and rubber may allow permeation to take place.

Table 7.3 (page 90) and Table 7.4 (page 91) respectively give some examples of the effects of chemical degradation and permeation.

The process of degradation is obvious and its consequences easy to see. The surface of the material will swell and lose its flexibility, making it liable to crack. As the material corrodes away, holes appear and the whole structure of the material is weakened. Discoloration is often an early sign that degradation is occurring.

In contrast, permeation is an invisible process, as the chemical is absorbed onto one surface of the material, diffuses through it and then evaporates from the other surface. Once a chemical has permeated into a material it can stay there for some time and be evaporated only slowly.

Permeation is different from *penetration*, which occurs when the material is imperfect and chemicals are able to pass through holes in it. Permeation takes place on a molecular level as molecules of a chemical diffuse through a material. The rate at which a chemical passes through a given sample of a material (its *permeation rate*) and the time it takes to travel from one side to the other (its *breakthrough time*) will depend upon several factors:

- a pure liquid will permeate faster than its vapour;
- a gas or vapour moving slowly across a surface will permeate more than one moving at high speed;
- breakthrough times decrease and permeation rates increase as the temperature rises;
- breakthrough times increase and permeation rates decrease as the material gets thicker;
- a positive pressure will increase permeation.

Table 7.3 Chemical resistance/degradation

Chemical	ABS	Br	Ni	Ny	PC	PE	PP	PVC	Rb	SS	Tf
Acetic acid, conc.	0	−	+	−	−	+	+	0	−	+	+
Acetic acid, dilute	+	−	+	−	+	+	+	+	−	+	+
Acetone	−	+	−	+	−	+	+	−	+	+	+
Acrylonitrile	nt	nt	nt	+	−	+	+	nt	nt	nt	+
Aniline	−	nt	−	0	−	+	+	−	−	+	+
Bleach solution	0	0	+	−	0	+	0	+	−	0	+
Butanol	nt	+	nt	+	+	+	0	nt	nt	nt	+
Caustic soda	+	+	+	+	−	+	+	+	+	+	+
Chloroform	−	+	−	0	−	0	−	−	−	+	0
Chromic acid	0	−	−	−	+	+	+	nt	−	−	+
Cyclohexanol	+	+	nt	+	0	+	0	nt	nt	+	+
Diesel oil	0	+	nt	+	+	+	+	+	nt	+	+
Ethanol	+	+	nt	+	+	+	+	+	+	+	0
Esters	−	+	+	+	nt	nt	+	+	−	+	+
Formaldehyde	+	+	+	+	+	+	+	+	+	+	+
Glycols	−	+	+	+	+	+	+	+	+	+	+
Hexane	0	+	nt	+	nt	+	+	nt	nt	+	+
Hydrochloric acid, concentrated	+	−	−	−	+	+	+	0	+	−	+
Hydrochloric acid, dilute	+	−	+	−	+	+	+	+	+	−	+
Hydrogen peroxide	+	−	−	−	−	+	+	+	−	+	+
Methanol	−	+	+	+	−	+	+	+	+	+	0
Methylene chloride	−	+	−	0	−	0	0	−	−	+	0
Nitric acid, conc.	0	−	−	−	−	0	0	−	−	−	+
Nitric acid, dilute	+	−	−	−	0	+	−	−	+	−	+
Nitrobenzene	−	+	nt	0	−	+	+	nt	−	+	+
Ozone	+	nt	−	−	+	0	−	+	−	+	+
Petroleum	0	+	+	+	−	+	+	nt	−	+	+
Sulphuric acid, concentrated	+	−	−	−	−	+	0	+	−	−	+
Sulphuric acid, dilute	+	−	−	−	+	+	+	+	−	−	+
Styrene	nt	nt	nt	+	−	+	+	nt	nt	+	+
Tetrahydrofuran	−	+	nt	+	−	0	0	nt	nt	+	−
Toluene	−	+	nt	+	−	0	0	+	nt	+	+
Trichloroethylene	−	+	+	0	−	0	0	−	−	+	+
Xylene	−	nt	nt	+	−	0	+	nt	nt	+	0

Key: ABS, acrylonitrile butadiene styrene; Br, brass; Ni, nitrile; Ny, nylon PC, polycarbonate; PE, polyethylene; PP, polypropylene; PVC, polyvinylchloride; Rb, rubber; SS, stainless steel; Tf, teflon or polytetrafluoroethylene (PTFE); +, resistant, only slight changes in mass or dimensions; nt, not tested; 0, limited resistance, some swelling and change in mass; −, not resistant.

Table 7.4 Permeation data for organic compounds*

Material	Breakthrough time in minutes for (thickness in mm) for:		
	Halogenated organics e.g., dichloromethane (methylene chloride)	**Aromatics e.g., toluene**	**Common solvents e.g., acetone (propanone)**
PVC	> 480 (5.0)	> 480	> 480
	65 (2.3)	102 (2.3)	180 (2.3)
	6 (1.4)	35 (1.4)	18 (1.4)
	2 (0.38)	6 (0.38)	4 (0.38)
Polycarbonate	> 480 (5.0)	> 480 (5.0)	> 480 (5.0)
Polypropylene	> 480 (5.0)	> 480 (5.0)	> 480 (5.0)
Natural rubber	18 (1.0)	26 (1.5)	95 (0.95)
	1 (0.31)	6 (0.6)	1 (1.4)
Nitrile	1 (0.34)	60 (0.6)	10 (0.75)
	6 (0.55)	20 (0.38)	3 (0.38)
Teflon	> 180 (0.35)	> 480 (0.48)	> 480 (0.28)
	64 (0.28)	> 480 (0.28)	
Viton	80 (0.30)	> 480 (0.30)	4 (0.25)
	60 (0.24)	> 480 (0.24)	
Laminates:			
e.g. Safety 4H	> 480 (0.07)	> 480 (0.07)	> 480 (0.07)
Du Pont Barricade	> 480 (0.5)	> 480 (0.5)	> 480 (0.5)

*These permeation data were measured at Respirex Testing Laboratory, 23/27 Endsleigh Road, Merstham, Surrey, RH1 3LX, and taken from Forsberg and Keith (1995) (see Appendix 2 for full reference).

Standard permeation tests are carried out according to BS EN 369 at 20–25°C with the material in continuous contact with the pure chemical for up to eight hours. In the case of a volatile organic solvent this will be the pure liquid. This, of course, represents a fairly unrealistic situation for most purposes but it does give a good comparison of the various materials and shows which ones are likely to be affected by permeation.

Table 7.3 gives examples of materials used in the construction of containers, half-suits and gloves and shows the effect of chemical degradation.

Table 7.4 shows that the thickness of the material is very important in preventing permeation. A sheet of 5-mm PVC used in the construction of an isolator will provide complete protection for 8 hours against an aggressive organic solvent (e.g., dichloromethane) but the same chemical will permeate through thin PVC or nitrile gloves in a few minutes. If protection against an aggressive chemical is required for hands and arms, then permeation data must be studied carefully to find a suitable material. Sometimes this means that a thicker material must be used to give the necessary protection, which will mean loss of tactile sensation. In some cases the use of double gloves will solve the problem.

The following typical thicknesses will be encountered for these materials:

• materials used in construction: 4–5 mm;
• tubing: 1–3 mm;
• gloves: 0.2–1.0 mm.

These examples show that the polymeric materials used in the construction of containment systems need to be considered carefully when the chemicals to be isolated are known. In particular, the fabric of the gloves used in isolators must be chosen carefully to ensure that operators are adequately protected.

Typical cleaning methods

It is important to remember that the ease with which typical containment devices can be cleaned varies according to design. For the purpose of this guide four grades of *cleanability* are defined as follows:

- C1 manual wash-down;
- C2 dismantle and wash-down with a solvent wipe;
- C3 decontaminated using Cleaning-in-Place (CIP) procedures;
- C4 CIP with contained dismantling (this will require the use of PPE, whose limitations are discussed in Chapter 9).

Containment device evaluation sheets

The control devices covered in this chapter have hitherto been described as generic types; improvements in the benefits they offer to specific transfer problems may well be possible with variations in the detail of the design. The remainder of this chapter comprises a set of containment device evaluation sheets, which illustrate typical devices that can be used to implement the four Containment Strategies considered in this guide (some of these devices are actually quite sophisticated systems). The containment device evaluation sheets summarize the operating principles, applicability and cleanability of each device.

The containment device evaluation sheets are grouped according to the Containment Strategy to which they apply. In some cases, the level of containment provided by a device depends on details of its construction or other devices with which it is used. For example, a split butterfly valve (page 87) or a rapid transfer port (page 85) in itself provides protection satisfying Containment Strategy 3 requirements but when used within an isolator it is suitable for applications requiring Containment Strategy 4.

The first aspect to consider in determining whether a control device is suitable for a given process is whether it can handle the required quantities of the material. Each containment device evaluation sheet, therefore, includes an indication as to whether the device is suitable for quantities of solids measured in grams, kilograms or tonnes or liquids measured in millilitres, litres or cubic metres.

The next aspect to consider is the dustiness of the solids or volatility of the liquids being handled, which in practice has only a minor impact on the suitability of the device.

Chapter 6 considered how to use the quantities handled, dustiness or volatility and the hazard band of the materials to determine which Containment Strategy is appropriate. If any of the special constraints are encountered (e.g., skin/eye irritation or radioactivity), the preliminary selection should be checked with a competent health and safety adviser.

Using the containment device evaluation sheets

In the 'Device Evaluation' section at the bottom of each sheet is a set of boxes that can be used to help the engineer whose task it is to assess the suitability of the device for a particular application. The boxes and instructions for using them are as follows:

- **Scale of operation** — This box has already been completed with ticks to show the quantities of materials for which the device is suitable. If the quantities used by the application in question are not ticked, then another device must be considered.
- **Hazard band** — The hazard band for the most hazardous material used in the application should be determined as described in Chapter 6 and entered in this box.
- **Required Containment Strategy (CS)** — Using the hazard band entered in the previous box, the required Containment Strategy should be determined, as described in Chapter 6 and with reference to Tables 6.4 to 6.7, and entered in this box.
- **Device suitability** — This box indicates the highest Containment Strategy for which the device is suitable. If the value determined in the previous step is higher than the value given in this box, then the device is not suitable and another device must be considered.
- **Over-riding constraints** — This table lists specific hazards that are presented by certain substances. Each material to be used in the device under consideration should be considered against these criteria and the appropriate 'Yes' or 'No' box ticked. If any 'Yes' boxes have been ticked, then the substance represents a serious hazard and expert advice should be sought on the additional precautions needed to use the device safely.

In general, the assessment should be based on the most hazardous chemical present in the process. However, if several hazardous materials are involved with different characteristics of dustiness, volatility or readiness to attract or react with moisture or other chemicals, then it may be better to evaluate the containment requirements for each material individually and choose a strategy that meets the needs of the worst case. In any case, if there is a change in the chemicals used in a process, the containment requirements should be reassessed.

List of containment device evaluation sheets

The following containment device evaluation sheets are included on the pages shown (the references before each title identify the Containment Strategy):

LOCAL EXHAUST HOODS (LEV)

What to look for:
- the enclosure should surround or be close to the dust/vapour source;
- flanging helps to focus air flow into the hood;
- the system should be self-cleaning;
- the hood should be easy to position;
- the system should be made of compatible materials (to minimize corrosion).

More detailed design information is given in the ACGIH guide (1998).

Diagram from American Conference of Governmental Industrial Hygienists (ACGIH®): Industrial Ventilation: A Manual of Recommended Practice, 22nd ed. Copyright 1995, Cincinnati, OH. Reprinted with permission.

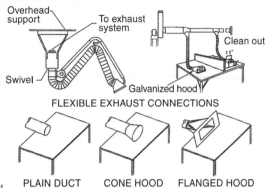

FLEXIBLE EXHAUST CONNECTIONS

PLAIN DUCT   CONE HOOD   FLANGED HOOD

Operating principle

Air from the vicinity of the source generating dusts or vapours is sucked into the hood, carrying with it the contaminants or pollutants. The air velocity is critical to the successful performance of the hood, as there is a minimum velocity — known as the *capture velocity* — required to divert the dust/vapour cloud into the hood. However, air velocity reduces greatly with distance, typically falling from 100% at the entrance to an open-ended duct connection to 20% at a distance from it equal to the diameter of the duct. Enclosing the dust/vapour source (e.g., within a booth) or adding flanging to the hood brings significant benefits.

Typical applications

Commonly used at locations where occasional dust/vapour emissions occur, e.g.:

- vessel loading points;
- small mixing tanks;
- other areas where incidental emissions may occur during normal operation.

Application with solvents/liquids

Commonly used to ventilate operations for dispensing liquids or mixing liquids with solids, where the volume of dust or vapour is easily controlled.

Cleanability: 1

Device evaluation

Overriding constraints: presence of these may necessitate re-evaluation

	Y	N
Skin/eye contact hazard		
Respiratory sensitizer		
Biohazard		
Carcinogen		
Explosive/flammable		
Radioactive		

Scale of operation		Hazard band	Required CS	Device suitability
Grams or ml				
Kilograms or litres	✓			
Tonnes or m³				

✓ + ☐ = ☐ ≤ CS2

95

HORIZONTAL LAMINAR-FLOW BOOTH

What to look for:
- internal surfaces must be smooth and parallel;
- there should be a perforated distribution plate at the back:
  — occupying the full height (H) and width (W) of the cross- section;
  — with a total open area of about 5% of the cross-sectional area;
- the filtration/exhaust system must be able to handle large air- flow volumes, i.e., around $0.5 \text{ m s}^{-1} \times W \times H$;
- circular aerofoils should be fitted around the entrance to the booth, to minimize turbulence;
- lighting should be provided through a clear panel in the ceiling and not interfere with the air flow.

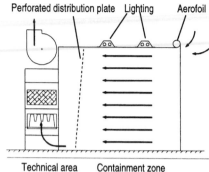

Operating principle

The chamber is open at the front and air is exhausted through the perforated plate at the rear. This sets up an inward air flow of uniform velocity that carries with it airborne contamination into the filtration/exhaust system. The filtration/exhaust system must be capable of handling large volumes of air, none of which is recirculated into the booth. However, the velocity of the air flow is low ($\sim 0.5 \text{ m s}^{-1}$) and hence *airborne material only* will be entrained.

The device will be disturbed by cross-draughts and poor operator work position. The operator should work sideways-on to the operation, standing as deep inside the booth as possible. Because of the large exhaust volumes employed, this design is not suitable for small work-rooms.

Typical applications

Commonly used for applications involving manual drum-to-process transfer, vessel charging, equipment dismantling, powder dispensing/subdivision or powder/liquid mixing.

Applications with solvents/liquids

Typical examples:

- vessel loading where a number of drums must be evacuated by dip pipe;
- open decanting of liquids;
- control of solvents during equipment cleaning/wash down.

Cleanability: 1

Device evaluation

Overriding constraints: presence of these may necessitate re-evaluation

	Y	N
Skin/eye contact hazard		
Respiratory sensitizer		
Biohazard		
Carcinogen		
Explosive/flammable		
Radioactive		

Scale of operation		Hazard band	Required CS	Device suitability
Grams or ml				
Kilograms or litres	✓	+ =	≤	CS2
Tonnes or m³	✓			

DRUM TRANSFER BOOTH/FUME CUPBOARD

What to look for:
- to minimize exhaust flow, only a small area of the access door should be open during operation — usually 10 to 20% of face area;
- doors should open on hinges for loading, sliding windows should be fitted to enable operator to access drum;
- a sump should be fitted to collect spilt liquids;
- the air-flow distribution plenum should be suitable for a low exhaust volume;
- lugs should be fitted to enable vessels to be earthed;
- cut-outs/pipe stubs for transfer of liquids should be within easy reach.

Operating principle

This type of device comprises an enclosure with hinged doors to allow drums to be loaded into it and a much smaller sliding window to enable the operator to access it whilst minimizing the volume of air to be exhausted. The quantity, Q, of air required per unit time while the operator is working on the drum depends on the air speed, V, (normally a constant at $0.5 \, \mathrm{m \, s^{-1}}$) and on the area, A, of the opening ($Q = V \times A$). Reducing the size of the access window will, therefore, minimize ventilation requirements. Pipe stubs, dip pipes and pump brackets should be appropriately placed for easy access through the small window so as to ensure ergonomic operation.

Typical applications

Used for drum-to-vessel transfer utilizing standard dip pipes. Drums must be loaded into the enclosure whilst closed and opened only after the booth doors have been closed, the access window has been opened and the ventilation system is operating.

Application with solvents/liquids

Not suitable for powders.

Cleanability: 1

Device evaluation

Overriding constraints: presence of these may necessitate re-evaluation

	Y	N
Skin/eye contact hazard		
Respiratory sensitizer		
Biohazard		
Carcinogen		
Explosive/flammable		
Radioactive		

Scale of operation		Hazard band	Required CS	Device suitability
Grams or ml	✓			
Kilograms or litres	✓	+ =	≤ CS2
Tonnes or m³				

BAG RIP AND TIP STATION/DRUM DUMP STATION

What to look for:
- the bag rest platform should be set at a comfortable height;
- the enclosure around the hopper and bag-cutting and dumping areas must be adequate to contain the dust liable to be generated;
- the perforated air-distribution plate should have an open area totalling about 5% of the cross-sectional area;
- breaker bars should be fitted to prevent the bag from falling into the hopper;
- arrangements for disposing of empty bags must be provided within the capture zone.

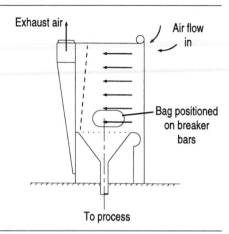

Operating principle

A device of this type comprises a miniature horizontal laminar flow booth with an inward flow of air from the room, covering a hopper for loading solids into a process. The bag is manually lifted and positioned onto the bag rest or breaker bars in the hopper. When the bag is in place, the operator uses a knife to open it — this knife should be retained in the enclosure (e.g., with a chain, or even built into it) to contain contamination within the booth. The empty bag and any other wastes are passed out (through a port in the side wall) into a collection sack, polythene bag or bag compactor.

Air flow will be disturbed by cross-draughts.

Typical applications

Used in bulk charging operations during the production of fine chemicals, paint and pharmaceuticals. Popular where large numbers of 25-kg paper sacks are to be loaded into the process. Bag lifters may be considered to aid bag handling. Automatic sack cutters may be desirable where their cleanability is acceptable.

Application with solvents/liquids

Not usually applicable for this design.

Cleanability: 1

Device evaluation

Overriding constraints: presence of these may necessitate re-evaluation

	Y	N
Skin/eye contact hazard		
Respiratory sensitizer		
Biohazard		
Carcinogen		
Explosive/flammable		
Radioactive		

Scale of operation		Hazard band	Required CS	Device suitability	
Grams or ml					
Kilograms or litres	✓	+	=	≤	CS2
Tonnes or m^3	✓				

PACKING SEAL (DRUMS WITH LINERS)

What to look for:
- inflation controls must be easy to reach and use;
- the diameters of the inflatable seal and the liner must be compatible;
- the packing head and any exhaust connections should be mounted onto the load platform itself;
- drum entry/exit routes require careful evaluation;
- the seal membrane must be compatible with product/solvents;
- it is desirable to filter or otherwise control the displaced air within the sealing head to minimize product loss.

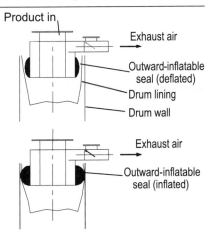

Operating principle

The drum to be filled rests on a weighing platform and its inner lining is sealed to the filler pipe by means of an outward-inflating seal device. A pair of concentric filler/exhaust pipes or a small-bore exhaust connection is used to load powder into the drum and capture air displaced by it.

When the drum is full, the powder feed valve will be closed (in many applications this happens automatically as soon as the weight of the drum and its contents reaches a preset level), after which the seal will deflate to allow closing of the liner.

This device should ideally be enclosed within a miniature horizontal laminar flow booth to capture any airborne contamination that may escape as the liner is closed and sealed.

Typical applications

Any process-unloading application where dry or wet solids are packed off into drums fitted with polyethylene or similar impervious liners. Where the number of drums is high, drum staging areas and run-off roller conveyors are common.

Application with solvents/liquids

Not usually applicable for this design.

Cleanability: 1

Device evaluation

Overriding constraints: presence of these may necessitate re-evaluation

	Y	N
Skin/eye contact hazard		
Respiratory sensitizer		
Biohazard		
Carcinogen		
Explosive/flammable		
Radioactive		

Scale of operation		Hazard band	Required CS	Device suitability
Grams or ml				
Kilograms or litres	✓	+	=	≤ CS2
Tonnes or m³	✓			

PACKING HEAD (DRUMS WITHOUT LINERS)

What to look for:
- inflation controls must be easy to reach and use;
- diameters of inflatable seal and drum must be compatible;
- control of displaced air exhaust or use of a filter within the sealing head is desirable to minimize product loss;
- the packing head and any exhaust connections should be mounted onto the scale platform itself;
- drum entry/exit routes require careful evaluation;
- the seal membrane must be compatible with product/ solvents.

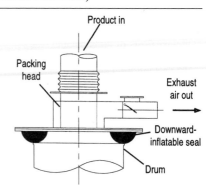

Operating principle

This application differs from the one on the previous page in that the drum to be filled has no liner bag. The packing head is normally suspended on a frame above the load platform and includes a downward-inflatable ring, which forms a seal when inflated against the rim of the drum. The weight of the packing head and support frame must therefore be taken into account in the measurements made. A flexible connection to the feed chute above the suspended packing head is also necessary. As in the example on page 99, this device must be enclosed within a horizontal laminar flow booth to capture any airborne contamination that may escape as the lid is closed.

Primary containment is achieved by a downward-inflating seal, which is kept in contact with the rim of the drum. When the drum is full, the feed valve will be closed and the seal will deflate. This will permit the drum to be pulled out away from the filling point so that the lid can be fitted in front of or to the side of the inflatable seal in such a way that the air flow in the booth will continue to capture any airborne material. As in the previous case, the closure of the valve and deflating of the seal when the drum is full can be automated.

Typical applications

Any process-unloading application where dry or wet solids are packed into drums without polyethylene or similar impervious liners. Where the number of drums is high, drum staging areas and run-off roller conveyors are common.

Application with solvents/liquids

Not usually applicable for this design.

Cleanability: 1

Device evaluation

Overriding constraints: presence of these may necessitate re-evaluation

	Y	N
Skin/eye contact hazard		
Respiratory sensitizer		
Biohazard		
Carcinogen		
Explosive/flammable		
Radioactive		

Scale of operation		Hazard band	Required CS	Device suitability
Grams or ml				
Kilograms or litres	✓	+	=	≤ CS2
Tonnes or m³	✓			

DOWN-FLOW BOOTH

What to look for:
- air should be delivered through a distribution system in the ceiling:
  - extending across the full width of the booth;
  - overhanging the working area in order to minimize the air drawn in from the surrounding room;
- air velocity must be uniform over the full width of the booth and depth of the working area;
- the exhaust volume is typically 3500–4000 m³ h⁻¹ per 1 m width;
- the internal walls should be smooth;
- the filter system must have a high dust-holding capacity and be backed up by HEPA;
- the booth must not be positioned in areas of high cross-winds;
- filters on the negative-pressure side of the fan are desirable.

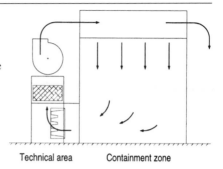

Technical area   Containment zone

Operating principle

Air is supplied through the ceiling and drawn out at the rear of the booth so that it flows over the entire work zone (typically at 0.5 m s⁻¹). High-velocity dust clouds are captured only in the rear section and hence the exhaust grilles should be carefully positioned in relation to dust/vapour sources. Dust capture at the edges of the containment zone is marginal.

In most applications a pre-filter followed by a High-Efficiency Particulate Arrestor (HEPA) is used to clean the exhausted air, which is then recirculated inside the booth; this is especially beneficial if the air outside the booth may be contaminated.

Typical applications

Versatile: used for powder subdivision, process loading, tray dryer loading and unloading, large scale powder pack-off systems and warehouse sampling.

Application with solvents/liquids

When none of the exhausted air is recirculated the design is suitable for handling solvents and or liquids.

Note: the need to supply and exhaust large volumes of air may quickly destabilize room/building pressure gradients unless fans are interlocked and operated with suitable flow control. Even when the booth is not in use it is desirable to maintain the air flow at a nominal level, e.g., 25% of its full operational speed, to prevent natural air flows from blowing back through the exhaust filters and releasing collected contaminants into the facility.

Cleanability: 1

Device evaluation

Overriding constraints: presence of these may necessitate re-evaluation

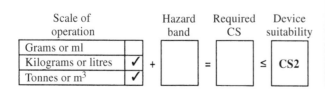

	Y	N
Skin/eye contact hazard		
Respiratory sensitizer		
Biohazard		
Carcinogen		
Explosive/flammable		
Radioactive		

Scale of operation — Hazard band + = Required CS ≤ Device suitability CS2

Scale of operation	
Grams or ml	
Kilograms or litres	✓
Tonnes or m³	✓

WORKSTATION DOWN-FLOW BOOTH

What to look for:
- the barrier must be ergonomically designed to ensure that required operations can be carried out;
- methods of transporting materials into or out of the containment zone need careful consideration;
- all other requirements listed on the previous page apply.

Operating principle

A device such as that illustrated is fitted inside a conventional down-flow booth (see previous page) to hold drums or other process vessels close to the 'exhaust wall' at the rear of the booth, where the exhaust system is most effective in removing any contaminants released. Such a device may be equipped with a barrier screen to keep operators at a safe distance from the transfer operation. Some form of lifting equipment may be required to load drums into the device. Arrangements of this kind can reduce the contamination inhaled by the operator by a factor of three compared with operations in a flow booth without such a guard. However, design of such devices needs to be carried out by specialists.

Typical applications

As per the standard down-flow booth but good for applications requiring improved levels of operator protection.

Application with solvents/liquids

As for the standard down-flow booth (previous page).

Cleanability: 1

Device evaluation

Overriding constraints: presence of these may necessitate re-evaluation

	Y	N
Skin/eye contact hazard		
Respiratory sensitizer		
Biohazard		
Carcinogen		
Explosive/flammable		
Radioactive		

Scale of operation		Hazard band	Required CS	Device suitability	
Grams or ml					
Kilograms or litres	✓	+	=	≤	**CS2**
Tonnes or m³	✓				

SHIELDED DRUM TRANSFER LANCE

What to look for:
- operation should be easy;
- an overhead support system should be provided for ergonomic operation;
- the flexible sheath should be reliable;
- the integrity of the ball valve/exhaust manifold is critical;
- the lance and the drum should be earthed to dissipate any static charge generated by the transfer.

Operating principle

This design takes the well-known vacuum dip pipe and places the part of the pipe that comes into contact with the liquid within a flexible sheath. The dip pipe passes through a close-fitting ball valve, which forms the base of an exhaust chamber. The dip pipe is entered into the drum by:

- removing the filling plug;
- positioning the dip pipe using an overhead support bracket;
- opening the ball valve, permitting the dip pipe to pass through the ball valve and penetrate into the drum.

The exhaust connection maintains the drum and dip pipe under negative pressure. At the end of liquid transfer, the dip pipe rises back to a position above the ball valve, which is then closed, thereby trapping any liquid residue and permitting the exhaust air flow to control any residual vapours.

Typical applications

Drum emptying, liquid transfer in inert environments, drum filling, etc.

Application with solvents/liquids

Confirm compatibility of dip pipe components against chemicals involved in the application.

Cleanability: 2

Device evaluation

Overriding constraints: presence of these may necessitate re-evaluation

Scale of operation		Hazard band	Required CS	Device suitability		Y	N
Grams or ml					Skin/eye contact hazard		
Kilograms or litres	✓				Respiratory sensitizer		
Tonnes or m³				≤ CS2	Biohazard		

Scale of operation [✓ Kilograms or litres] + [Hazard band] = [Required CS] ≤ CS2 [Device suitability]

	Y	N
Skin/eye contact hazard		
Respiratory sensitizer		
Biohazard		
Carcinogen		
Explosive/flammable		
Radioactive		

103

AIR-WASHED BUTTERFLY VALVE

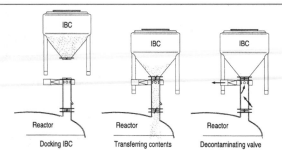

| Docking IBC | Transferring contents | Decontaminating valve |

What to look for:
- surface finish should be smooth;
- all joints between the device and the source and receiving containers should be effectively sealed;
- the device should be fitted with both supply and exhaust air flows;
- the sequence of operations should be controlled automatically.

Operating principle

This device comprises a chute, with its own air supply/exhaust, closed at the bottom by a butterfly valve. It is built onto a vessel (e.g., a reactor) and used to contain the transfer of powders or solids into that vessel from a mobile container (e.g., an Intermediate Bulk Container (IBC), which is also closed at the bottom by a butterfly valve. The butterfly valves are interlocked so that they may be opened only if the chute is attached to a compatible mobile container.

Any residual powder or dust left in the chute after the transfer is complete is removed by closing both butterfly valves and blasting high-velocity air (or an inert gas) into the chute. Meanwhile, the exhaust air flow maintains negative pressure in the chamber relative to the working area and draws the contaminated air away for dispersal outside. When this is complete the mobile container may be removed.

Typical applications

Often seen at IBC discharge stations where powdered materials will be allowed to fall from an IBC into the process vessel.

Application with solvents/liquids

Not common in liquid applications.

Cleanability: 1

Device evaluation

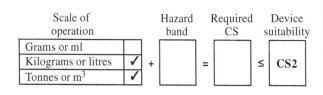

Overriding constraints: presence of these may necessitate re-evaluation

	Y	N
Skin/eye contact hazard		
Respiratory sensitizer		
Biohazard		
Carcinogen		
Explosive/flammable		
Radioactive		

Scale of operation		Hazard band	Required CS	Device suitability	
Grams or ml		+	=	≤	CS2
Kilograms or litres	✓				
Tonnes or m³	✓				

SPLIT BUTTERFLY POWDER VALVE

What to look for:
- the device must include a reliable docking/interlocking mechanism;
- seals must be effective in minimizing powder loss when the valve is closed;
- compensators must permit slight adjustment of the accuracies that will be tolerated during docking;
- air/nitrogen blast and exhaust couplings are desirable;
- automation should be considered for larger sizes.

Operating principle

Used to transfer powder from a portable container into another vessel. Transfer ports fitted to the powder container and the receiving vessel comprise matching halves of the split butterfly mechanism; these are closed (locked closed in some designs) to prevent powder loss from the container during transportation and orientation into the docking position.

The two ports are locked together with the butterfly discs closed. The locking of the two ports is usually achieved by using bayonet fixings or a rotating sleeve. Once secure, the two valve discs open together to permit direct transfer of powder from the container to the vessel. Accurate machining or seals alleviates powder contamination of the faces of the butterfly discs, which, thus, have only traces of contamination when separated after transfer. Regular attention and maintenance are essential to ensure reliability and ease of use.

Further details of the operation of this type of device are given on page 9.

Typical applications

These systems are traditionally used in kg or tonne applications where hygienic bulk powder transfer between hoppers, containers, isolators and IBCs or other vessels is required.

Application with solvents/liquids

Not suitable.

Cleanability: 2

Device evaluation

Overriding constraints: presence of these may necessitate re-evaluation

	Y	N
Skin/eye contact hazard		
Respiratory sensitizer		
Biohazard		
Carcinogen		
Explosive/flammable		
Radioactive		

Scale of operation	
Grams or ml	
Kilograms or litres	✓
Tonnes or m³	✓

Hazard band + [ ] Required CS = [ ] Device suitability ≤ CS3

SPLIT BUTTERFLY LIQUID COUPLING

What to look for:
- the coupling should be interlocked so as to prevent inadvertent opening of the fluid line;
- the coupling should be easy to use;
- the couplings should be compatible with the chemicals to be transferred;
- the discs should seal together effectively;
- the bore of the tube should be smooth to minimize the drop in pressure across the coupling.

Identical discs in the coupler and adapter are brought together to connect the assembly

PATENTED

The discs join and swivel together to permit flow.

Operating principle

Similar concept to the split butterfly powder valve (page 105) but used to transfer fluids from a vessel into a pipeline. The vessel outlet or inlet and the hose will have matching couplings comprising a locking flange and interlocking valve discs. When the hose and vessel outlet are locked together, the locking mechanism releases the flow control handle. This permits both discs to open in unison, permitting fluid transfer. After transfer, reversal of the above steps permits hygienic de-coupling.

Typical applications

Flexible hose used to connect pipelines to vessels.

Application with solvents/liquids

Review manufacturers' data to ensure that the method and materials of construction of both the valve and the coupling are compatible with the liquids to be transferred.

Cleanability: 2

Device evaluation

Overriding constraints: presence of these may necessitate re-evaluation

	Y	N
Skin/eye contact hazard		
Respiratory sensitizer		
Biohazard		
Carcinogen		
Explosive/flammable		
Radioactive		

Scale of operation		Hazard band	Required CS	Device suitability	
Grams or ml					
Kilograms or litres	✓	+	=	≤	CS3
Tonnes or m³	✓				

SAFE-CHANGE AIR FILTER HOUSING

What to look for:
- construction must be robust;
- the access cover must be easily removed over a double-groove spigot or transfer port;
- an effective cam lift mechanism and knife edge seal must be fitted;
- filters are pulled out by means of a handle;
- this handle and the cam lift handle must be compatible with working through a polythene overbag;
- filters used must be free of sharp corners or edges that may damage the overbag;
- housing may need to be made of stainless steel or alloy;
- a 'DIN test' device for testing the gasket seal the seal is desirable.

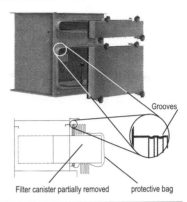

Filter canister partially removed   protective bag

Operating principle

The air filter is mounted in a special canister, which can slide into or out of the filter housing. When inside the housing, the filter is compressed against a seal bulkhead by a cam lift mechanism. This mechanism eliminates the requirement for handwheels or multiple clamps to seal the filter gasket to the housing. The cam mechanism is tightened and released by a simple handle that requires typically 90° of rotation. This simple tool- free sealing mechanism permits the entry port of the canister to be covered by a large polythene bag. The spigot that holds this bag has a double-groove feature similar to that described on the previous page that, with the correct procedure, will permit removal and disposal of a contaminated filter through the bag. A clean filter can be loaded through another bag, over which the access cover is then fitted. With careful heat-sealing and cutting of the bags, similar to the process described on the previous page, the remnants of the old bag can be removed under cover of a part of the new one without any need for the operator to touch the contaminated filter, the interior of the housing or a contaminated bag.

Typical applications

These filter housings, grouped or single, are necessary in any application where the collected dust could cause harmful contamination of the operator or service area.

Application with solvents/liquids

Not suitable.

Cleanability: 2

Device evaluation

Overriding constraints: presence of these may necessitate re-evaluation

	Y	N
Skin/eye contact hazard		
Respiratory sensitizer		
Biohazard		
Carcinogen		
Explosive/flammable		
Radioactive		

Scale of operation		Hazard band	Required CS	Device suitability
Grams or ml				
Kilograms or litres	✓ +	=	≤	CS3
Tonnes or m³				

GLOVE-BAG CONTAINMENT UNIT

What to look for:

- construction should be of good quality with reinforced base sheet and glove-port fittings;
- it must be possible to change gloves and air filters without compromising the integrity of the containment provided;
- the bag must be supported by a frame fitted with a tensioning system;
- access/exit routes must not compromise containment;
- advice should be sought from the supplier on correct design and use.

Operating principle

Glove-bags are flexible barrier isolators with limited service life. This technology was developed in the USA for occasional service applications. The bag fabrication is usually clear-welded PVC or urethane with glove-ports and access ports welded in during manufacture. A support frame to hold the glove- bag open and venting filters are usually provided.

In operation the clean glove-bag is sealed or fixed over the clean process equipment item. Hazardous materials are entered into the unit in sealed bags introduced via bag entry tubes or zippered ports. All powder handling is conducted via the glove-ports with the bag remaining closed and under negative pressure. The bag must be cleaned and de-contaminated before it can be removed. Bags can last from 3 to 6 months depending on the care taken of them.

Typical applications

Common in pilot plant and laboratory-scale applications where usage will not be continuous. Commonly seen as enclosures around small-scale powder- processing plants such as dryers, mixers and compression machines. May be adapted for powder weighing or loading powders into process vessels.

Application with solvents/liquids

Not common, as solvents may attack the bag envelope.

Cleanability: 2/3

Device evaluation

Overriding constraints: presence of these may necessitate re-evaluation

	Y	N
Skin/eye contact hazard		
Respiratory sensitizer		
Biohazard		
Carcinogen		
Explosive/flammable		
Radioactive		

Scale of operation		Hazard band	Required CS	Device suitability	
Grams or ml	✓				
Kilograms or litres	✓	+ $\Box$ = $\Box$		≤	**CS3**
Tonnes or m³					

RIGID-SHELL GLOVE-BOX ISOLATOR

What to look for:

- internal surfaces must be constructed so as to be smooth and coved in order to avoid crevices where contamination could build up;
- the windows that form the operator interface should be inclined;
- the ergonomics must suit all operators involved;
- the glove ports should permit glove replacement without compromising integrity of containment;
- safe-change air/dust filtration elements are essential;
- controls and/or alarms must be fitted to indicate (un)satisfactory negative pressure gradients;
- material entry/exit must not breach containment;
- the enclosure should be self-draining if it is likely to be washed down;
- a bag-out system is desirable for disposal of waste material.

Operating principle

A compact rigid enclosure is fitted around a process or operation, with operator interaction limited to glove-ports at strategic positions. Because of the difficulty of lifting containers much above 5 kg via glove ports, lifting devices are often incorporated in designs where drums are to be accessed.

Typical applications

May be applied to most bulk and medium-scale operations requiring transfer of powders or solids, e.g.:

- process loading;
- powder weighing;
- process off-loading (via packing head);
- milling and micronizing.

Application with solvents/liquids

Occasionally used, but glove resistance to solvent/liquid may be limiting factor.

Cleanability: 2/3

Device evaluation

Overriding constraints: presence of these may necessitate re-evaluation

	Y	N
Skin/eye contact hazard		
Respiratory sensitizer		
Biohazard		
Carcinogen		
Explosive/flammable		
Radioactive		

Scale of operation		Hazard band	Required CS	Device suitability
Grams or ml				
Kilograms or litres	✓	+	=	≤ CS3
Tonnes or m³	✓			

HALF-SUIT ISOLATOR

What to look for:
- the suit must be of robust construction to minimize the risk that tears or cracks will permit operator contamination;
- the visor must provide good visibility;
- it must be possible to replace glove sleeves or cuffs without compromising the integrity of the containment provided (the suit will outlast the gloves);
- the half-suit must include a ventilation system to cool the operator and provide respirable air;
- a hanging/support system for the suit will help the operator to enter it.

Operating principle

A 'half-suit', i.e., a flexible over-suit designed to cover the operator's head, arms and torso, is securely attached to a port at the base of an isolator. The isolator can comprise a rigid shell, as with a glove-box (page 113), or welded PVC soft walls, as used for a glove-bag (page 112). The base of the isolator will usually be formed from a fully-welded stainless steel tray with a large oval port to suit the dimensions of the half-suit. Operator access is achieved by climbing into the half-suit, which is ventilated for comfort.

Typical applications

The half-suit isolator is often used where a large degree of operator movement is needed, for example sterility testing, freeze-dryer unloading or milling and micronizing applications. For industrial applications, the isolator enclosure will generally be of a rigid-shell construction.

Application with solvents/liquids

Not suitable, as solvents may attack the material of the suit.

Cleanability: 2/3

Device evaluation

Overriding constraints: presence of these may necessitate re-evaluation

	Y	N
Skin/eye contact hazard		
Respiratory sensitizer		
Biohazard		
Carcinogen		
Explosive/flammable		
Radioactive		

Scale of operation		Hazard band	Required CS	Device suitability
Grams or ml	✓			
Kilograms or litres	✓ +	=	≤	CS3
Tonnes or m³				

LINED FIBC 'BIG BAG'

What to look for:

- powders/solids within the FIBC are contained within a polythene liner;
- the liner neck should permit attachment of a suitable seal during FIBC filling;
- the FIBC outlet should have an extended discharge tube with a powder tie-off at the underside of the bag — this permits attachment of a clean discharge tube to the receiving vessel prior to powder transfer;
- some designs use inflation of the liner bag to aid product discharge;
- hygienic removal of the empty liner is desirable.

Operating principle

Within a Flexible Intermediate Bulk Container (FIBC) is fitted a hygienic liner to contain the bulk material. The liner usually features an extended neck and out-feed tube of 200–400 mm diameter that permits filling and emptying with minimal operator contact with the solids. The FIBC can be used in conjunction with rigid-shell isolators to provide higher levels of containment than that offered by an unlined FIBC.

This arrangement is a specialist design and requires careful operating processes.

Typical applications

Bulk transfer of free-flowing solids into and out of the process train. Useful for one-way trips where IBC technology would not be commercially viable.

Application with solvents/liquids

Not suitable.

Cleanability

Not applicable.

Device evaluation

Overriding constraints: presence of these may necessitate re-evaluation

Scale of operation		Hazard band	Required CS	Device suitability		Overriding constraint	Y	N
						Skin/eye contact hazard		
						Respiratory sensitizer		
Grams or ml		+	=	≤	CS3	Biohazard		
Kilograms or litres						Carcinogen		
Tonnes or m³	✓					Explosive/flammable		
						Radioactive		

115

RAPID TRANSFER PORT (DPTE®
OR ALPHA-BETA PORT)

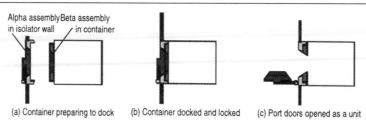

| (a) Container preparing to dock | (b) Container docked and locked | (c) Port doors opened as a unit |

What to look for:
- quality of manufacture should be assured by ordering only from a proven supplier;
- the seal must be reliable whilst minimizing the area of contact between the seals and possible contamination;
- size, fitting and materials of construction must be compatible with containers/chemicals used;
- an interlock should prevent the port door from opening unless the container is docked;
- compatibility considerations should decide whether a suitable industry-standard product is available or a special unit must be made;
- A lifting device may be required for larger-sized units (above 300 mm in diameter).

Operating principle

These ports enable hazardous materials to be transferred from a container into an isolator and vice versa without loss of containment. Both the container carrying the hazardous material and the wall of the isolator must have compatible port designs: the *alpha assembly*, built into the wall of the isolator, and the matching *beta assembly*, which closes off the container. Both assemblies comprise a frame holding a circular door, which cannot be opened unless a matching assembly is docked in place, in which case the two doors open together as one.

The container is offered up to the locked port door in the isolator wall as shown in figure, (a), above. The alpha and beta assemblies are docked together, (b), and rotated between 45° and 90° (depending on the design). This locks the port and the container together and also interlocks the removable discs that form the closure pieces in each half. Only when fully interlocked can the discs be lifted or hinged out of the way, (c), usually by the operator working through glove ports in the isolator.

Typical applications

Most commonly used where occasional transfer of solids into an isolator needs to be undertaken without any breach in containment. May also be used to transfer powders into or out of containers, provided that the port contact surfaces are protected from contamination by the powder.

Note: the port cannot control pressure or material flow, so additional valving may be necessary if such control is required.

Application with solvents/liquids

Not commonly used with solvents.

Cleanability: 2

Device evaluation

Overriding constraints: presence of these may necessitate re-evaluation

Scale of operation			Hazard band	Required CS	Device suitability
Grams or ml	✓	+		=	≤ CS3
Kilograms or litres	✓				
Tonnes or m³					

	Y	N
Skin/eye contact hazard		
Respiratory sensitizer		
Biohazard		
Carcinogen		
Explosive/flammable		
Radioactive		

ENCLOSED MICRONIZING PLANT

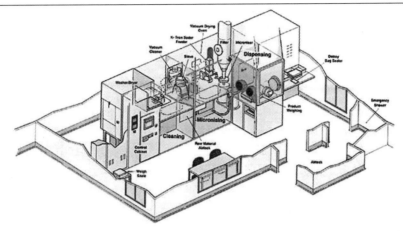

What to look for:
- no open powder transfer may be permitted, even within the enclosure;
- material must be loaded via a contained transfer coupling;
- material must be packed off via a contained transfer or into a continuous-liner system;
- internal finishes must be easily cleaned;
- all filters must be easy to replace safely;
- the enclosed system must be self-draining if CIP is used;
- an operator station fitted with half-suits may provide ergonomic benefits.

Operating principle

Material handling within the enclosure is minimized by using a vacuum transfer system to carry the hazardous material from the loading station through the micronizer and into the receiving container. Material must be transferred into or out of the enclosed system via contained transfer couplings and must not be manually scooped, even under the protection of a glove-box.

Typical applications

Milling/micronizing of highly hazardous materials such as pharmaceutical actives and research compounds.

Application with solvents/liquids

Not applicable.

Cleanability: 3

Device evaluation

Overriding constraints: presence of these may necessitate re-evaluation

Scale of operation		Hazard band	Required CS	Device suitability		Y	N
					Skin/eye contact hazard		
					Respiratory sensitizer		
Grams or ml					Biohazard		
Kilograms or litres	✓ +	=	≤	CS4	Carcinogen		
Tonnes or m³					Explosive/flammable		
					Radioactive		

AUTOMATED FILLING LINE

What to look for:
- all hazardous materials must be delivered to filling device by pipeline;
- filling devices must be fully automatic;
- operations to prepare, fill or seal the container must all be totally enclosed;
- containers must be free of contamination when they leave the enclosure.

Operating principle

The term 'automated filling line' covers a wide range of applications and designs. In the context of this guide it refers to lines with usual operator input minimized or eliminated completely. The hazardous material will be typically in the form of a liquid. This will be directed into the isolator by rigid pipeline connections. Containers will be prepared (and sterilized if required) automatically before being filled with the product. After the containers are filled they will be sealed, again under automated control. It may also be necessary to decontaminate the outside of the containers before they leave the filling line enclosure. In some cases, the receiving container may be created, filled and sealed all within the isolated system.

Such a system would be suitable for Containment Strategy 5 applications if operator input to the process were eliminated completely. However, the example illustrated involves some operations using glove-ports and so must be rated as Containment Strategy 4.

Typical applications

Automated filling lines are limited to volume production applications. Those requiring containment will include parenteral pharmaceuticals, aerosols carrying pharmaceutical actives and agro-chemicals.

Application with solvents/liquids

Commonly used with liquids, creams and occasionally powders.

Cleanability: 4

Device evaluation

Overriding constraints: presence of these may necessitate re-evaluation

Scale of operation		Hazard band	Required CS	Device suitability			Y	N
Grams or ml	✓					Skin/eye contact hazard		
Kilograms or litres	✓	+	=	≤	CS4	Respiratory sensitizer		
Tonnes or m^3						Biohazard		
						Carcinogen		
						Explosive/flammable		
						Radioactive		

TOTALLY AUTOMATED PROCESS

What to look for:
- the operator must have no contact at all with process materials;
- the system must be fully automated — operated by robots under remote control.

Operating principle

Materials for which Containment Strategy 5 is required are so hazardous that no person may be allowed to come into contact with them. The processes that use them must therefore be totally enclosed and fully automated.

The design of such systems is a highly specialized activity. They are available only from a limited number of vendors. Specialist designers must be consulted.

Typical applications

As systems of this kind need to be designed individually for specific applications, there is no such thing as a typical example.

Application with solvents/liquids

Specialist designs only.

Cleanability: 4

Device evaluation

Overriding constraints: presence of these may necessitate re-evaluation

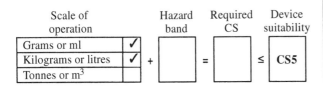

Scale of operation		Hazard band	Required CS	Device suitability
Grams or ml	✓			
Kilograms or litres	✓	+ =	≤	CS5
Tonnes or m³				

	Y	N
Skin/eye contact hazard		
Respiratory sensitizer		
Biohazard		
Carcinogen		
Explosive/flammable		
Radioactive		

Control of wastes and emissions

Purpose

The purpose of this chapter is to consider briefly the regulatory issues around BATNEEC, the environmental impact of the containment of hazardous substances and the safe containment and disposal of materials and wastes contaminated by them. Explosions have the potential to cause especially severe harm, both to individuals and to the environment, and this chapter gives extensive consideration to preventing and controlling them.

Contents

BATNEEC

The UK COSHH Regulations and EU Directives state that the use of closed systems for the containment of hazardous materials is the preferred approach for minimizing emissions and operator exposure. Three key points arising from this are that:

- the cost versus the benefit of the closed system needs to be established;
- closed systems handling hazardous materials minimize but still usually produce contaminated wastes;
- various types of closed systems are available to choose from.

Best Available Technique Not Entailing Excessive Cost (BATNEEC) is a very powerful element of the international standards and EU Directives relating to environmental protection and of the UK **Environmental Protection Act 1990** (EPA), which is derived from them. In this context the term *technique* includes technology but is also intended to cover operational factors, which are important for containment systems. Currently, the selection of a solution based on BATNEEC is left to the user.

All processes prescribed under Part 1 of the EPA are subject to BATNEEC requirements. In general terms, what is BATNEEC for one process is likely to be BATNEEC for a comparable process. However, the applicant can appeal to the Secretary of State to decide what is BATNEEC for an individual process, in which case variable factors such as configuration, size and other individual characteristics of the process will be considered. The courts could be the final arbiters.

For these reasons, it will, in practice, be necessary to have a general working definition of BATNEEC for the guidance of inspectors in the field, the operators of *scheduled processes* and the Secretary of State in determining appeals or in issuing directions to the Chief Inspector and to local authorities.

In reducing the emissions to the lowest practicable amount, account needs to be taken of:

- local conditions and circumstances in relation to:
  — the process;
  — the environment;
- the current state of knowledge;
- financial implications in relation to:
  — capital expenditure;
  — revenue cost.

The constituent parts of BATNEEC will now be described in turn.

BAT

The term 'best available technique' embraces both the process used and how that process is operated. It includes not only the hardware components but also matters such as staff numbers, training, supervision and operating procedures.

'Available' means that the technique is procurable by the owner. It does not imply that the technique/technology is in general use.

'Best' must be taken to mean most effective in preventing pollution, minimizing it or rendering it harmless. It implies that the effectiveness of the technology has been adequately demonstrated.

NEEC

'Not entailing excessive cost' needs to be taken in two contexts, depending on whether it is applied to new or existing processes.

The presumption will be that 'best available techniques' will be used, but that presumption can be modified by economic considerations where it can be shown that the costs of applying 'best available techniques' would be excessive in relation to the environmental protection to be achieved (or the viability of the company).

If, for instance, one technology reduces emissions by 90% and another by 95% but at four times the cost, which one is BATNEEC? If the extra 5% pollution were not critical in the overall sense, then the second technology would entail excessive cost. However, if the extra emissions were still particularly dangerous, it may be proper to judge that the additional cost is not excessive. It remains up to the engineer to make a considered judgement supported by the business case.

All containment techniques selected within BATNEEC should protect personnel by keeping exposures below the limits for the materials handled. The usual major BATNEEC decision for process containment is the balance between the cost and possible operational restrictiveness of closed systems and the perceived low cost of open handling of materials by operators wearing Personal Protective Equipment (PPE) such as full air suits. Those responsible for making this comparison must be aware of hidden building costs associated with the areas in which open material handling takes place; these costs will include provision of:

- the containment suite;
- changing and showering areas;
- the heating, ventilation, filtration and air-circulating system.

These extra costs should be included in the evaluation of any proposed open- handling solution to enable a realistic comparison to be made against the cost of a fully isolated system.

As exposure limits are reduced, operators' working practices and work patterns may become increasingly important, especially if they have been used to justify simple containment. The introduction of duration into an evaluation imposes a monitoring burden upon management. The location of the sampling position is also of critical importance in making a realistic assessment of what the operator is liable to breathe.

Consequently, these factors have to be taken into account in determining the protection technique to be adopted to meet the requirements of BATNEEC and the COSHH Regulations, which seek to minimize the wearing of PPE.

However, whether an isolator system or an open system with air-suited operators is used, hazardous emissions and wastes will arise.

122

Wastes from containment systems

It is reasonable to assume that any item or material that leaves a containment system could be contaminated with hazardous materials and hence pose a danger to the environment. It is necessary, therefore, to evaluate all people, material and process streams fully to ensure that no uncontrolled release occurs. The waste streams would include:

- empty packaging;
- empty process containers;
- contaminated filters;
- contaminated clothing;
- exhaust gases containing hazardous vapours or particulates;
- contaminated drainage and cleaning liquids.

The levels of contamination and the approaches taken depend on the containment system selected for the problem and the potential hazards related to the process media handled.

Empty packaging

Hazardous solid materials will typically arrive at plants in bags or in drums with liners. Hazardous solid materials manufactured and transferred between stages within plants, or between different plants within the same company, may be stored:

- in liners in drums;
- directly inside plastic containers;
- in semi-bulk containers;
- in the case of the most hazardous or potent materials, in containers with Rapid Transfer Ports (RTPs) or equivalent devices.

Handling empty sacks and liners can create dust clouds, which require control as described below.

Containment by air-flow devices – Containment Strategy 2

Within down-flow booths, it is usual to bag up empty sacks and liners into one sack or drum, which can be decontaminated on the outside and then safely disposed of according to company procedures.

More careful attention is required when tipping operations are carried out within laminar flow booths. It is possible to provide a simple bag-out chute through which contaminated packaging can be loaded into sealed bags that remain clean on the outside. Such a chute can form an integral part of the booth so that the empty bag or liner can be put into it directly without being removed from the air flow. It is also possible to include a compactor within this chute to reduce the volume of waste packaging that needs to be disposed of.

Containment in isolators – Containment Strategies 3 to 5

Where bags, liners or sacks are emptied within a closed isolator, a bag-out device is needed to remove the waste packaging in a similar manner to that described above for air-flow devices but with a level of transfer protection appropriate to the Containment Strategy. Where material is handled in disposable bags within and between isolators fitted with RTP container couplings, more careful attention must be paid to the disposal of the final waste packaging. One of the isolators should be designed to have a safe disposal route out of the system.

In some designs for equipment to contain highly toxic materials, the flexible container has a plastic RTP fitted, which, like the container itself, is relatively inexpensive and can be safely destroyed together with the container after use.

Empty solid process containers

Rigid closed containers, such as plastic drums or larger Intermediate Bulk Containers (IBCs), have to be cleaned in such a way that no people are exposed to concentrations of the hazardous materials inside them above their relevant exposure limits.

Depending on the design of the closed container, and particularly the type of valve it uses for filling and emptying, proprietary Cleaning-in-Place (CIP) packages are available, which may be suitable for cleaning it. These wash the container both inside and outside, rinse it and then dry it using hot air. The containers should be designed according to hygienic principles so that they can be easily cleaned. Procedures for achieving the required decontamination should be drawn up, validated and routinely followed.

Where simpler containers are used for transferring water-washable solid materials, wash booths should be installed in the plant at locations near their point of use. A hot and cold water utility station is required together with a contained drainage system. The operator should wear suitable protective clothing.

Empty liquid drums

It is good safety practice for standard 200-litre chemical drums to be detoxified at the point of use or within the plant and not sent out of the works still containing residues of chemicals that may be hazardous. This can be achieved in wash booths, but where hazardous water-reactive chemicals are to be destroyed, proprietary drum-cleaning devices are available or simple systems can be constructed. In these, the relevant washing/decontamination fluid is prepared and then poured into and drained out of the drum, with the process repeated as necessary until the residues are reduced to a safe level.

Contaminated filter elements and media

Many processes use filters to remove impurities or particulates from the liquids or sludges they generate. This leads to a further problem: the filters need to be cleaned in a way that does not cause further contamination. (A similar problem, that of cleaning the filters used to remove hazardous particulates from exhaust air streams, will be covered on page 125.)

The hazardous waste materials that have to be removed in this way will usually be wet and so should pose a minimal dust problem. There may, however, be hazardous vapours from the solvents used to clean them. Hence, a controlled and contained arrangement for removing chemicals from the filter is necessary, depending on the degree of hazard. Typical options include:

- for filter elements and cakes, locating small filters in small air-flow booths (Containment Strategy 2) with direct bag-out facilities;

- for more hazardous materials, locating filters inside isolators (Containment Strategy 3 or 4) with closed in/out component transfers;
- cake discharge from larger filters direct to closed drums, or other containers.

Whilst filters may be contaminated with hazardous waste solids, the filter cake may also be wet with flammable organic solvents. In this case it is usual practice to flush the cake or filter with water to remove the flammable solvents before removing the cake or the whole filter element.

Contaminated clothing

In theory, since operators should not be exposed to hazardous substances, their normal clothing should not be contaminated during manual or other operations involving the transfer of solids or liquids.

In closed isolator systems (Containment Strategies 3–5) this will indeed be the case. However, in localized capture systems (Containment Strategy 2), dusts will be generated during open tipping within the air-flow booth and so parts of operator's clothing could well become contaminated. Over time and depending on the material, this could then spread contamination as the operator leaves the booth and does other duties in the plant. Where this may cause a problem, the operator should, before entering the booth, don appropriate clothing or other PPE such as:

- disposable paper overalls;
- a full air-suit.

In the latter case, it is good practice to install a shower directly next to the booth to wash off contamination from the suit before the operator passes in to the general plant area, or ensure that the suit is removed in an area directly adjacent to the booth. In the former case, arrangements must also be made for removing and disposing of the contaminated overall in such a way as to contain the contamination.

Airborne dust may spread to the environment from rooms where hazardous dusts are openly handled in simple air-flow devices (Containment Strategy 2) and so prevention of contamination from this source may be a problem for some manufacturers handling substances that require air-suit operation. The principle for such devices has been to enclose the booth inside a small room to prevent the general spread of dust to other plant areas.

Exhaust gases

Many containment devices have gas flows to remove dust from the working environment. With Containment Strategy 2 designs, large volumes of air are used to capture dusts liberated during open operations and substantially remove them from the operator's breathing zone. For Containment Strategies 3 to 5, smaller flows of air or inert gases are required to maintain low dust levels within the closed isolators. When flammable solvents or hazardous solids are used in the process, it is common for isolators to be purged with nitrogen, which is either recirculated via the isolator filter or purged through the isolator in a 'once through' mode to the atmosphere via High-Efficiency Particulate Arrestors (HEPAs). HEPAs should be protected by suitable pre-filters to prolong service life and reliably maintain a consistent air-

125

flow. Where the risk of HEPA failure is deemed unacceptable then a second, back-up HEPA may be installed.

Where possible, all primary air-flow filter systems should be located as near to the point of contamination as possible to prevent contamination of ductwork, e.g.:

- at the exit of localized capture booths (Containment Strategy 2);
- as integral parts of isolators (Containment Strategies 3–5).

For isolators handling substances of extreme potency or hazard (mainly Containment Strategy 5), two in-line HEPA filters are recommended, one providing a back-up to the other.

These filter elements should all be installed so that they can be changed from within the isolator and removed through bag-out or RTP devices in suitable closed containers. This duty to conduct maintenance within the isolator has important implications for its design, requiring consideration of:

- access to the isolator filter;
- the ergonomics of change operation;
- filter change with back-up systems in place;
- protection of air filters from liquids used in CIP or other isolator cleaning processes.

Contaminated drainage and cleaning liquids

The safe disposal of liquids drained from processes containing hazardous materials or those used to clean process equipment or containment devices is an important consideration in the operation of any plant. It is self-evidently not good practice to discharge liquids containing hazardous substances from the containment zone into open drainage outside the plant, where solids could dry out and cause a problem.

Consequently, all drainage should be contained and the drained liquids disposed of in a safe manner that does not spread hazards into the plant. Safe disposal could comprise one of the following:

- closed drainage to the process drainage system for central collection and safe treatment to detoxify waste and discharge;
- closed drainage to a separate segregated system where the hazardous substance can be specifically treated;
- a filter on the drainage pipe from the containment device to remove suspended solids.

All streams leaving the containment device and its associated services, such as shower wastes from cleaning air-suits, should be evaluated for potential hazards and, where necessary, treated according to one of the methods described above.

Explosion prevention and control

In general, it is recommended that containment systems should be provided wherever they are needed and that they should be closed whenever it is practical and financially affordable to make them so.

However, any process capable of generating a flammable atmosphere that could be ignited presents a threat to the integrity of a closed containment system unless protective measures are taken. Such a process might involve dry solids, which could cause a potentially explosive dust cloud under some circumstances, or other materials in flammable atmospheres.

Examples include:

- an isolator around a mill;
- larger-scale open tipping of solids into flammable solvents inside a contained system where static charges may build up.

Obviously all electrical equipment, instrumentation and lighting must be specified and installed to meet the appropriate explosion-proof and surface temperature rating for the chemicals handled.

Solutions to this problem include:

- provision of an inert nitrogen atmosphere inside the isolator;
- designing the isolator shell and its components to:
  - withstand a 10-bar internal pressure (or more, depending on the explosive behaviour of the dust cloud);
  - prevent rupture if an explosion should occur;
- provision of a weak rupture panel which vents via a large filter or scrubbing device to atmosphere.

For the latter case, it is necessary to estimate the expected volumetric flow rate and volume of a released gas and design the secondary containment system to prevent a release to atmosphere. Released gases are often vented into a larger catch-tank to reduce their pressure before they are passed to the secondary containment device.

Some of the issues involved in explosion control and venting inside containment systems are described below. Further information can be found in the IChemE design guide edited by Barton (2001).

Explosion control inside isolated systems

Explosion control and suppression inside a contained environment is an important consideration in the design of closed process equipment in which dust or vapour explosions could occur. Several solutions can be employed by those designing such a system.

These options generally fall into one of five categories, whose merits will be discussed later in this chapter:

- *zoning:* dividing the workplace into *zones* according to the fire risks present and ensuring that all equipment is approved for use in the zone where it is required — this eliminates a significant source of ignition;
- *full containment:* designing the system to withstand the full explosive pressure that could occur in the event of an explosion;
- *limited venting:* designing the system for venting into a confined area in the event of an explosion;

127

- *inerting:* designing the system to inert the process, i.e., to incorporate substances (usually an inert gas) that will be present at all times and that will not affect the process but by displacement of air (oxygen) will prevent an explosive mixture from forming (this approach may be used in conjunction with other methods);
- *explosion suppression:* designing the system so that it will introduce substances to suppress the explosion as soon as it occurs, to prevent pressure build-up and subsequent damage — this is usually regarded as a last resort.

General considerations

Clearly, as with any process, there is a risk of an explosion within an isolated system. However, by reducing the size of the production area and confining it by barrier techniques, the risk of explosion has been automatically reduced.

The use of an isolator as a physical barrier between the process and the general environment reduces the likelihood that an explosive hazard from outside the enclosure will cause an accidental explosion.

Therefore the designer, when considering explosion control, will need to concentrate only on:

- the process itself;
- the area within the confines of the isolator;
- any vent points that may be present in the unit.

There are, though, some issues that, when applied to barrier systems, require more attention. These are discussed in detail in later sections.

The designer and design team should firstly ensure, at the earliest possible stage, that any parts of the process that may be a source of ignition are designed out.

Issues that should be considered are:

- elimination of hot surfaces;
- static bonding and cross-bonding;
- friction generation;
- lighting (exposed);
- radio frequency;
- the nature of any gas, dust or vapour present;
- lower explosive (flammable) limits;
- potential for dust clouds (and likely concentrations);
- sources of ignition (e.g., by mechanical contact);
- release point into the isolator;
- vaporization rates;
- isolator air change rates;
- vapour density — where a build-up of vapour or dust may occur;
- flash points of any vapours present;
- auto-ignition temperatures of any vapours or gases present;
- minimum energy that will ignite any mixture present;

- presence of light alloys and dust;
- exothermic reactions;
- reactions from any mixtures of chemicals present.

Once these have been considered and the design completed, it must be subjected to rigorous review as part of the **_Design Qualification_** (DQ) procedures described in Chapter 10. This review would also cover the project's own safety risk assessment to identify anything else that could form an ignition source in the process or isolator.

By completing these steps the designer will have a rational and documented set of evidence to support any decision made with regard to the explosion control method selected.

Hazard assessment

Figure 8.1 (pages 130–131) summarizes the hazard assessment process that leads to the identification of any potential explosion hazard and the control system required to overcome or minimize that hazard.

Zoning

A **_Hazardous Area Classification_** (HAC) analysis will normally be carried out for all processes involving flammable vapours and dusts. This is done by:

- examining the process together with:
  — the containment equipment it uses;
  — the space it occupies;
- determining whether there is sufficient vapour, dust or gas to generate an explosion if a source of ignition were present (as described in BS 6713 Parts 1–4).

The HAC process then demands that the design team remove from the explosion risk area all equipment that may generate a source of ignition.

This can be done in two ways:

- eliminating, where possible, electrical equipment in that part of the space;
- using flameproof or intrinsically safe equipment suitable for the hazardous zone classification in that part of the isolator system — the equipment will be surface temperature-rated and be unable to generate an ignition source even under the worst conditions.

BS EN 60079-10 defines a classification system for working areas and equipment suitable for use in them. Further guidance on the selection of suitable electrical equipment can be found in BS EN 50015 to 50021 and BS 5501-8 and in Cox, Lees and Ang (1990).

Completion of a HAC analysis is the minimum requirement if there is a risk of explosion. It is good practice to remove electrical equipment from within isolators, e.g., by using external lights or motors, even if this equipment is rated as flameproof or intrinsically safe. Methods used to limit the potential damage to a process system if an explosion were to occur are described below.

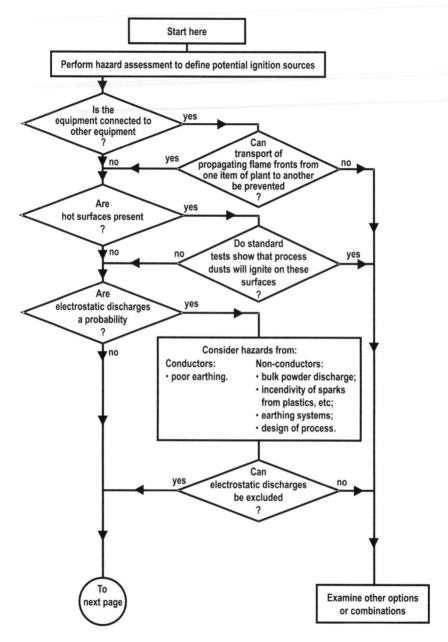

Figure 8.1 Hazard assessment process

Full containment of explosive pressure

This is the ideal route if the substances contained are highly toxic and particularly unstable and therefore cannot be released to the atmosphere. However, in this case ergonomic issues must be fully addressed at the design stage, as many of the component parts will be permanently inaccessible or heavy by design in order to meet the required high-pressure rating.

130

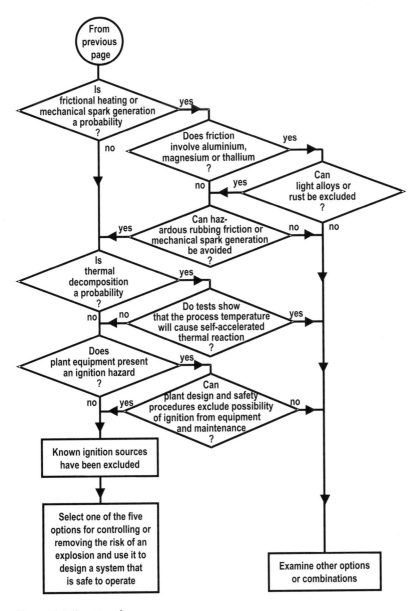

Figure 8.1 (*continued*)

This method is of particular relevance to milling systems. However, some consideration must be given to preventing the spread of flames to other adjacent equipment.

Thus there are two closely related aspects to consider:

- equipment configuration;
- control of explosions.

Equipment configuration

If there is a significant explosion risk, due care must be taken with the design of any multi-chamber or isolated systems, whose stages cascade from one to another, to prevent an explosion from being propagated through the interlinked system. Ideally, the explosion will be contained in a controlled area so as to prevent any emission of toxic product from the system and minimize the risk of product loss or damage to any other part of the process.

In a complex isolator system that consists of several interconnected working spaces, care must be taken to separate these spaces and, if possible, restrict the free volume of any powder or liquid delivery systems to a minimum. The possible use of ganged slide valves in particular may be one alternative to consider. Rotary valves can also act as flame pressure barriers between interconnecting systems in which solids are transferred.

Care must be taken with interlinked ventilation systems to include pressure- rated fire dampers where necessary to prevent movement of flame fronts through the system in the event of a fire. These dampers must be pressure-rated to withstand explosions and so are very expensive.

Control of explosions

Before any of these items are designed into the system, it is important to:

- establish the maximum explosion pressures that could be generated by any product that may be processed in the contained system;
- ensure that the design pressure is capable of withstanding this.

It may be prudent in some cases to have these parameters evaluated externally when dealing with a new compound.

It should be remembered that as the flame-front spreads through the ducts and chambers of a system it generates ever-increasing pressures. Unless great care is taken in the design to contain it, this spread of fire can quickly generate pressures in excess of 10 times the normal operating pressure.

In a simple configuration it may be possible to calculate the maximum pressure that a system may generate, but the more complicated a system becomes, the more difficult this is. Typically, an explosion in a single vessel may produce a pressure peak between 7 and 10 times the normal operating pressure.

In designing for explosion containment, two options are available:

(a) design to withstand the maximum explosion pressure with no equipment distortion;
(b) design to a lower pressure but accept that distortion will occur in the event of an explosion provided that any distortion will not breach the containment provided by the process equipment.

Option (a) is acceptable and produces a system that is very reliable. However, there is a significant penalty to pay with regard to construction and cost. To build in this way requires both significant testing and insurance inspections as well as the high-pressure rating of many of the working parts of the system. This will almost inevitably create many ergonomic issues, especially related to manual handling, which will themselves need to be addressed.

Option (b) should be used only when the likelihood of explosion is very low, as lost production time and the cost of replacement could be prohibitive. This option does, however, reduce some of the ergonomic issues caused by option (a).

It should be remembered that the basis for any design to control an explosion must be recorded at the Design Qualification (DQ) stage of a project, as described in Chapter 10, and form part of the system documentation.

Figure 8.2 (page 134) illustrates the decisions to be made when considering full explosion containment for a system. Further information can be found in the IChemE guide edited by Pilkington (2000).

Limited venting

If limited venting of an explosion from an isolator is acceptable, great care must be taken at the design stage.

This should be considered only if substances (including raw materials and intermediate or end products) of low toxicity will be vented into a ***highly controlled area***, i.e., one to which no personnel have access and where complete emergency action plans and spillage procedures exist.

The first question the design team must ask is whether it is acceptable to allow any spillage of these substances. If it is, then they must undertake a full project risk assessment, which includes considering the environmental issues that may be caused in the event of venting an explosion. If at this stage free venting is not acceptable, then an alternative option must be sought as discussed below.

It is important to establish the explosive characteristics of the substances being processed, so that a suitable venting route from the closed plant can be designed. This explosion escape route *must* be designed such that it is away from all personnel and that it will withstand the maximum explosion pressure generated from any of the substances being processed.

Consideration must be given to the size and strength of the vent area and their relationship to those of its associated isolation equipment. Time should be taken to ensure that the plant is configured in such a way as to reduce the risk of explosion, for example by considering:

- static bonding;
- non-sparking equipment;
- use of electrical devices suitable for use where an explosive atmosphere is likely to occur during normal or abnormal operations.

It is important that the vent route and vent itself should provide sufficient relief for the explosion so that a dangerous back-pressure into the process equipment is not possible. The isolation unit and vent route must ideally not be connected to any other plant. If, for example, the system to be vented includes interconnecting process vessels or ductwork, then the effects of a release on the rest of the plant must not affect their integrity and where a vented system cannot achieve this, an alternative method of explosion control must be selected.

The plant area into which the explosion is to be vented *must* be highly controlled with no access permitted during plant operating times. There should also be spillage equipment available adjacent to this area.

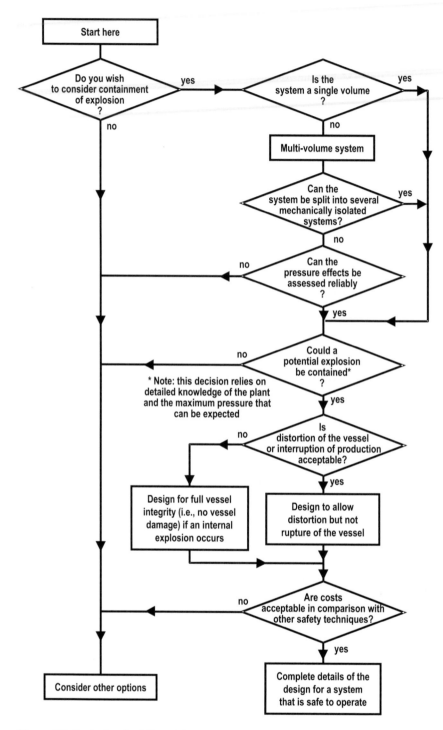

Figure 8.2 Designing for full containment

This process should be fully documented with full HAC and a HAZOP carried out to ensure that all exhaust routes will withstand the maximum explosion pressure and that the effects of the exhaust on the vent areas have been fully examined.

Where free venting of hazardous material is not allowed, excess gases from the isolator must be exhausted into a much larger closed container or an efficient vent scrubber/filter system employed. Where such a system of venting into another, larger, container is used, pressure-volume calculations must be carried out in order to determine the design pressure for the system. Any scrubbers and filters used in the vent stream should obviously not cause a blockage when in use and should be of sufficient area to prevent back-pressures exceeding the design pressure of the isolator.

Figure 8.3 (pages 136–137) illustrates the decisions to be made when considering limited venting for a process.

Inerting

Inerting is normally considered as the basic option for explosion control. It must be viewed in relation to both cost and operability. In many instances, inerting of isolator systems is carried out with the other protective measures, particularly for solids of a more hazardous nature.

However, it does afford the designer the opportunity to install a system that is ergonomic and cost-effective to build as there is no requirement for heavy bulkheads, seals or pressure-rated vessels, pipes and ducts, over and above those that the process requires.

This freedom does come at a price, which is reflected in the operating costs of the system, i.e.:

- cost of the inert gas being used;
- leakage from the inerting system;
- risk of accidental asphyxiation to anyone who enters the isolator when it is in operation;
- cost of the monitoring system that ensures that the process is inert at all process times;
- disposal of the inerting gas, which is liable to be contaminated.

Nitrogen is normally used as the inerting gas as it is cheap and widely available, it reacts with hardly any substances and, in particular, it does not react with pharmaceuticals.

Figure 8.4 (page 138) illustrates the decisions to be made when considering inerting for a process.

Explosion suppression

Introducing a substance to suppress an explosion can be a practical alternative if the substance being processed in a closed system is unstable in some way or if the risk of an explosion is relatively high. By using such a suppression system, the weight of any component parts can be significantly reduced since high pressures are in theory prevented from occurring. However, for pharmaceutical substances there are cGMP issues that must be considered, arising from the risk of accidental detonation of suppressant and consequent waste or contamination of the product.

For this reason, suppression systems are rarely used in pharmaceutical plants in closed systems where final products are being handled.

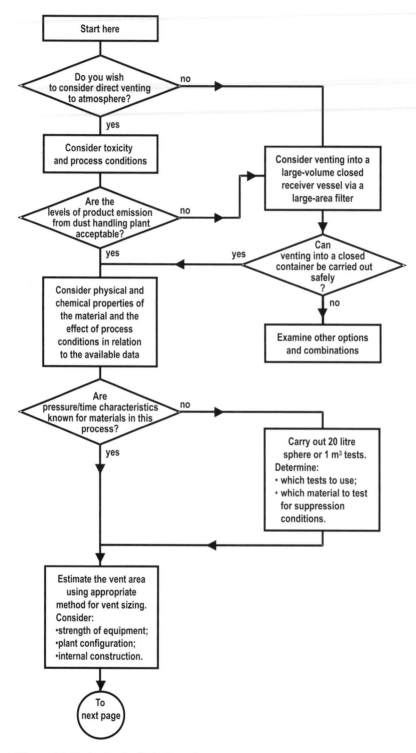

Figure 8.3 Designing for limited venting

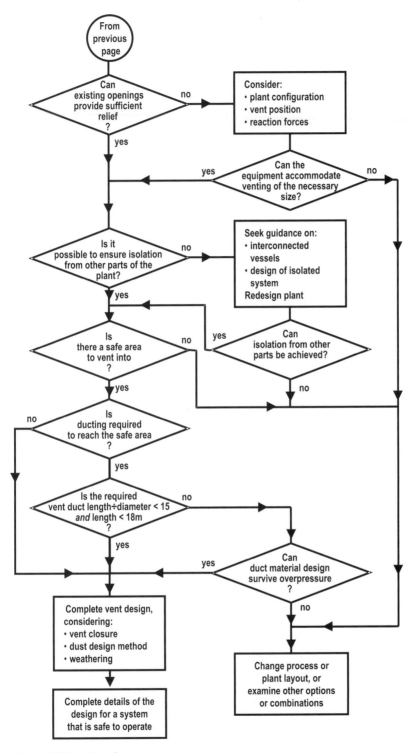

Figure 8.3 (*continued*)

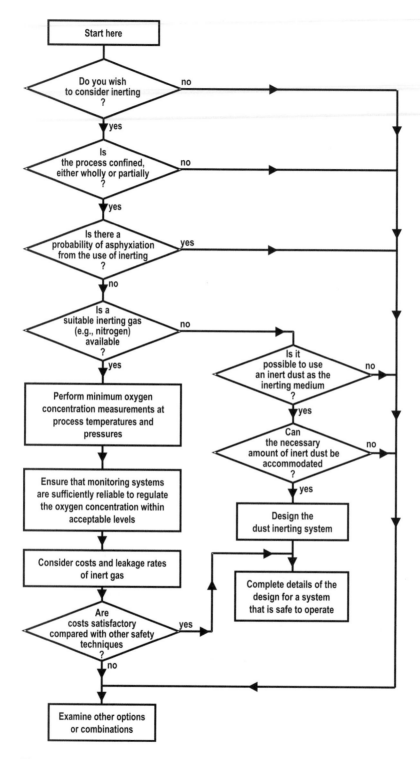

Figure 8.4 Designing for inerting

The suppressant to be used must be chosen carefully to ensure that it will not react with any chemicals in the process.

The design team must therefore take into account whether the process and equipment proposed are inherently suited to an explosion suppression system.

During the DQ stage, the design team must look at the physical and chemical properties of the substances used and consider what would happen if they were to come into contact with each other in the event of a leakage or explosion. This will also help them to decide whether the process and equipment proposed are inherently suited to an explosion suppression system.

For the suppression system to work efficiently it is important at the design stage to have available the pressure/time characteristics for the substance in the event of a deflagration. This will enable the manufacturer of the suppression system to determine accurately the type of system required and where the injection points should be. Access to these areas for recharging and maintenance must be considered.

It may also be worth considering the use of an explosion pressure monitor, which would capture the initial pressure wave data so that it is possible to determine the exact cause of any system activation or false detonation.

If a suppression system is selected then the design team must ensure that the isolation unit is capable of containing all the suppressant that may be injected into any of the explosion pathways. They must also address, during the design stage, all issues relating to cleaning of suppressant from the equipment or isolator.

The design team must also consider any interconnections to other isolators or other parts of the process to make sure that these are adequately suppressed or that there is no route for any flame or pressure wave to pass down.

Figure 8.5 (pages 141–142) illustrates the decisions to be made when considering full explosion suppression for a process.

Oxygen levels

Where explosion risks are finite for the substances being used, oxygen analysis is carried out to demonstrate that the atmosphere is safe before the process is started up. In practical terms, the design team will need to take into account the efficiency of the oxygen monitoring system, the explosive characteristics of the dust, vapour or gas and the size of the isolator. The isolator size becomes important, as a gas change rate during inerting of 20 volume change rates per hour may be used. This may have a significant bearing on cost, particularly if the purge philosophy requires that nitrogen should pass straight through the system without circulating and the vented gas has to be treated by other devices.

In order to provide a good margin of safety before processing can begin, typical oxygen levels after initial purging must usually be maintained below about 10%–20% of the level, known as the *limiting oxygen concentration*, above which the atmosphere becomes explosive. If the isolator or closed system is of large size with multiple vent exit points, then a multi-point monitoring system may need to be employed.

Special care must be taken to ensure that the sampling point of the oxygen analyser's sensor does not become blocked — it is strongly recommended that two sampling points should be used to obviate this danger. The location of the sensors must be carefully chosen to avoid placing them in 'dead pockets', i.e., areas out of the way of normal gas circulation where oxygen levels may be atypically low, and thereby give a false sense of security. However, additional sensors may be needed in dead pockets if atypically high oxygen concentrations are liable to accumulate there. It is better still if such dead pockets can be designed out of the system.

Conclusion

In addition to the detailed requirements of the appropriate Containment Strategy, the designer of any containment system must consider certain other safety aspects that are relevant to all such systems. These include adequate provision for disposing of the substances whose hazards are contained and for the prevention and containment of explosions, as described in this chapter. Another important consideration is maintenance of the containment system to ensure its continuing effectiveness, as discussed in the next chapter.

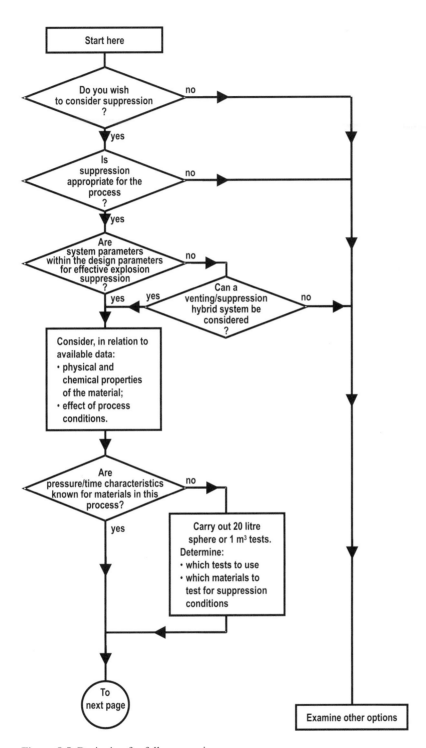

Figure 8.5 Designing for full suppression

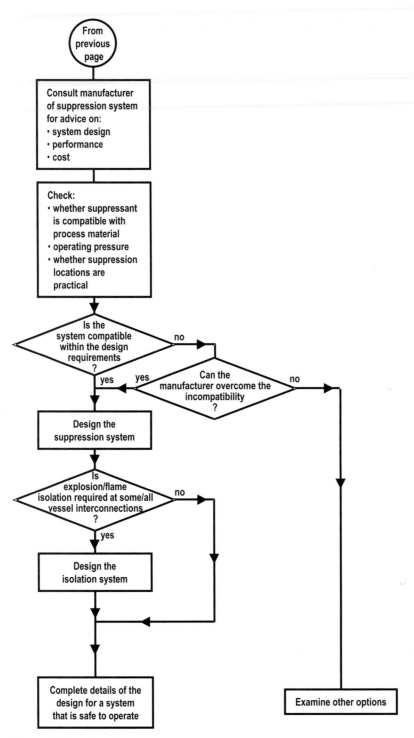

Figure 8.5 (*continued*)

Operation and maintenance of containment devices

Purpose

The purpose of this chapter is to provide guidance on the operational and maintenance requirements of the various types of containment devices. As Chapter 6 has shown, correct operation (via operator training) and correct maintenance are essential components of the selected Containment Strategy.

Contents

Introduction

Containment devices, irrespective of the Containment Strategy for which they are appropriate, are effective only when their operation (including the materials they contain) and their maintenance are in line with the designer's intentions.

In addition to critical design and ergonomic issues, the operation and maintenance requirements of any control device must be considered at an early stage, ideally during the initial and detailed design phases of the project. Where the containment device is to be specially built for a specific application, the designer must take these requirements into account. If the device is a commercially available product, the supplier should be consulted as to its suitability in the light of these operational and maintenance requirements. As part of the procurement programme for the containment device, time should be allocated for liaison between the users and suppliers of the selected devices as shown in Table 9.1.

Table 9.1 Interface between users and suppliers

Group	Stage of procurement programme
User group (operators) engineering group	Development of detailed *User Requirement Specification* (URS) outlining the objectives for the project and the budget and operations required to complete it.
Maintenance group	Review of URS to ensure maintainability.
User group (operations)	Review of ergonomic model or similar device.
User group	Signing-off equipment during validation stages (see Chapter 10).
User group	Involvement in initial operational trials.
Maintenance group	Involvement with: • commissioning and filter testing procedures; • ergonomic model.

Ensuring that the users accept the selected containment device and understand how to use it makes good practical sense. However, there are also legal obligations embodied in the UK's **Control of Substances Hazardous to Heath Regulations 1999** (COSHH), as described in Chapter 2.

Training for correct use of containment devices

The choice of Containment Strategy, as described in Chapter 6, will determine the level of operator training and maintenance attention required. The higher the Containment Strategy the more intense the training needed for operation and maintenance activities.

Typical operator training topics for the various Containment Strategies are listed in Table 9.2 (pages 145–146). Some of these may appear to be more relevant to installing or accepting a plant than to training operators to use it. Nevertheless, they should be considered during the preparation of any training course as they may identify hidden problems that the operators could encounter when using the devices.

Use of the training topic guide

It should be remembered that these typical operator training requirements are *guidelines only*. Each specific application/operation requiring a containment device *must be evaluated in its*

Table 9.2 Typical operator interface and training requirements (topic guide)

Containment Strategy 2: Local exhaust ventilation	Containment Strategy 3: Open handling within isolator	Containment Strategy 4: Closed handling within isolator	Containment Strategy 5: Robotic handling, contained system
Confirm that the operations and processes involved, including all container sizes to be used, match the design brief. Explain theory and provide tutorial on device operation. Use air-flow pattern visualization (smoke testing).	Confirm that the operations and processes involved, including all container sizes to be used, match the design brief. Explain theory and provide tutorial on device operation. Discuss handling techniques and air flows.	Confirm that the operations and processes involved, including all container sizes to be used, match the design brief. Explain theory and provide tutorial on device operation. Discuss handling and product transfer methods.	Confirm that the operations and processes involved, including all container sizes to be used, match the design brief. Review principles of total isolation of process operation. Discuss operating systems.
Evaluate materials/ container flow through workspace.	Evaluate materials/ container flow through workspace.	Evaluate material and container entry/exit routes.	Evaluate requirements of process: materials flow into and out of process.
Review emergency procedures. Dry run with non-hazardous materials: • monitor real-time exposure.	Review emergency procedures. Dry run with non-hazardous materials: • monitor ergonomic acceptability; • monitor exposure at entry/exit ports and operator station.	Review emergency procedures. Dry run with non-hazardous materials: • monitor ergonomic acceptability; • monitor exposure at entry/exit ports and operator station.	Review emergency procedures. Conduct dry run loading non-hazardous materials via control room: • monitor process for containment loss, if any.
Evaluate real-time exposure and discuss peak exposures. Modify work routine if required to reduce peak exposures.	Evaluate operating cycle and discuss improvements or changes. Modify routine if necessary.	Evaluate operating cycle and verify that direct connection systems maintain the required level of containment. Evaluate/demonstrate direct connection systems for: • ease of use; • containment integrity.	Evaluate operating cycle and verify that direct connection systems maintain the required level of containment. Evaluate/demonstrate direct connection systems for: • ease of use; • containment integrity.

(continued)

145

Containment Strategy 2: Local exhaust ventilation	Containment Strategy 3: Open handling within isolator	Containment Strategy 4: Closed handling within isolator	Containment Strategy 5: Robotic handling, contained system
Carry out exposure monitoring during typical loading cycle.	Set up containment proving test with suitable test material and carry out exposure monitoring during typical cycle.	Set up containment proving test with suitable test material and carry out exposure monitoring during typical cycle.	Set up containment proving test with suitable test material. Demonstrate that break-out of contamination from the contained system satisfies accepted criteria for safe exposure levels.
	Measure airborne and surface contamination. If unsatisfactory, modify device/work routine.	Measure airborne and surface contamination. If unsatisfactory, modify device/work routine.	
Document and record approved work procedures/routine.	Document and record approved work procedures/routine.	Document and record approved work procedures/routine.	Document and record approved work procedures/routine.
Observe approved routine whilst carrying out personal exposure monitoring to establish bench-mark exposure for task/operation.	Observe approved routine whilst carrying out personal exposure monitoring to establish bench-mark exposure for task/operation.	Observe approved routine whilst carrying out personal exposure monitoring to establish bench-mark exposure for task/operation.	Observe approved routine whilst carrying out personal exposure monitoring to establish bench-mark exposure for task/operation.
Seek approval of safety department for overall design.	Seek approval of safety department for overall design.	Seek approval of safety department for overall design.	Seek approval of safety department for overall design.

own right to ensure that the training provided meets the requirements of the appropriate legislation or company standards. For example, Regulation 8 of the COSHH Regulations states:

'(1) Every employer who provides any control measure, personal protective equipment or other thing or facility pursuant to these Regulations shall take all reasonable steps to ensure that it is properly used or applied as the case may be.'

Maintenance of containment devices

Regulation 9 of the COSHH Regulations states:

'(1) Every employer who provides any control measure to meet the requirements of regulation 7 shall ensure that it is maintained in an efficient state, in efficient working

order and in good repair and, in the case of personal protective equipment, in a clean condition.

(2) Where engineering controls are provided to meet the requirements of regulation 7, the employer shall ensure that thorough examinations and tests of those engineering controls are carried out:

(a) in the case of local exhaust ventilation plant, at least once every 14 months, or for local exhaust ventilation plant used in conjunction with a process specified in column 1 of Schedule 4, at not more than the interval specified in the corresponding entry in column 2 of that Schedule;

(b) in any other case, at suitable intervals.

(3) Where respiratory protective equipment (other than disposable respiratory protective equipment) is provided to meet the requirements of regulation 7, the employer shall ensure that at suitable intervals thorough examinations and, where appropriate, tests of that equipment are carried out.

(4) Every employer shall keep a suitable record of the examinations and tests carried out in pursuance of paragraphs (2) and (3) and of any repairs carried out as a result of those examinations and tests, and that record or a suitable summary thereof shall be kept available for at least 5 years from the date on which it was made.'

This places a legal obligation on employers to ensure that the various containment devices used in their workplaces are examined and tested regularly and maintained in good condition. Testing and maintenance must be carried out by a 'competent person', who will not only understand the operation of each device but will also make the necessary arrangements to:

- decontaminate the device;
- protect the surrounding environment from contamination;
- dispose of contaminated parts, filters, etc., safely.

As with other areas in this guide, the level of examination and testing will vary with the Containment Strategy within which the device is operating.

Table 9.3 provides key points to be covered in examination, maintenance and testing procedures. It is important to note that these are *key* points: actual requirements may well change from design to design.

Record keeping

The records of the above tests and any other information generated as a result must be retained and made available for inspection by the Health and Safety Executive (HSE) for a period of five years.

Weekly visual inspection

The General COSHH ACOP, L5 recommends that visual inspection of the equipment and containment device should be carried out at appropriate intervals. This may be set up with a simple operator check sheet to cover areas such as air flow and filter condition gauges and the condition of isolator gloves and half-suits.

Table 9.3 Typical requirements for examination, maintenance and testing

Containment Strategy 2	Containment Strategy 3	Containment Strategy 4	Containment Strategy 5
Checks of: • name/tag number of device to be examined; • date of last service; • hours run since last service.	Checks of: • name/tag number of device to be examined; • date of last service; • hours run since last service.	Checks of: • name/tag number of device to be examined; • date of last service; • hours run since last service.	Checks of: • name or tag number of device to be examined; • date of last service; • hours run since last service.
Comparison of actual performance with the design specification, with regard to: • capture velocities; • smoke containment test; • static pressure behind exhaust grille; • filter pressure drop.	Comparison of actual performance with the design specification, with regard to: • capture velocities at inlet/outlet ports; • if relevant, capture velocities at glove port; • effect of glove failure.	Comparison of actual performance with the design specification.	Test procedures specific to design.
Filter integrity testing: Dioctylphthalate (DOP) or similar where High-efficiency Particulate Arrestor (HEPA) used.	Filter integrity testing: DOP or similar.	Filter integrity testing: DOP or similar.	
Particle count (if required).	Particle count (if required). Glove leak/perforation test. Isolator shell pressure test/leak test.	Particle count (if required). Glove leak/perforation test. Isolator shell pressure test/leak test.	
Overall damage inspection. Fan condition, speed, RPM, etc.	Overall damage inspection. Fan condition, speed, RPM, etc. Operation of control systems.	Overall damage inspection. Fan condition, speed, RPM, etc. Verification of correct operation of direct connection devices.	
Motor and controls; also Programmable Logic Controller (PLC) sequence, where fitted.	Motor and controls; also PLC sequence, where fitted.	Motor and controls; also PLC sequence, where fitted.	

Spare parts

Naturally, once the correct examination and test programme has been planned, arrangements must be made to procure the spare parts necessary for operating and maintenance. These will include filters, safe-change bags, gloves, seals, gaskets, etc. and should be held in stock in accordance with the supplier's recommendations. Similarly, arrangements need to be put into place for safe disposal or incineration of contaminated waste, such as filters.

Operator protection during maintenance procedures

Wherever possible, the containment device should be designed so as to minimize the risk to maintenance staff during routine maintenance and cleaning. For example, air-flow devices that handle hazardous powders should be equipped with air filters having 'safe change' bag-in/bag-out change- over. Glove-box isolators will almost certainly have contamination-free change-over filters and will often incorporate Cleaning-in-Place (CIP) washing systems for decontaminating chambers prior to entry by personnel (e.g., for adjustment or maintenance).

Should there be any risk of contamination of maintenance personnel, the employer must provide appropriate Personal Protective Equipment (PPE). In the event that PPE is needed, consideration must also be given to containing any contamination fall-out, e.g., by sealing off and depressurizing the working area around the containment device and providing a decontamination system for PPE (e.g., suit shower and changing area). Portable facilities, such as polythene tents, are available from a number of sources. The need for PPE for maintenance staff should be considered when the overall maintenance programme is planned, to ensure that no personnel will be contaminated throughout the maintenance operations.

Selection of the appropriate PPE system

Types of protective clothing

In the event of a breakdown in a containment system, or if testing or maintenance work has to be carried out in an area where toxic materials could be exposed, operators will need to wear protective clothing. The equipment needed will be of one of four types, which are described in the European standards BS EN 465, BS EN 466-1 and BS EN 467.

Type 3 or 4

If the risk is associated with occasional splashes from corrosive and/or non-volatile chemicals, then a chemical splash suit will be sufficient when worn with protective boots, gloves and visor. This would give protection against all the commonly encountered acids and alkalis and the suit could consist of any of the following:

- separate jacket and trousers worn with a separate hood (Type 4);
- a one-piece suit worn with a separate hood (Type 4);
- a one-piece suit having an integral hood (Type 3).

Type 2

If the acids are very concentrated and fuming, then a complete coverall splash suit with integral visor, boots and gloves should be used. This would be worn over personal Respiratory Protective Equipment (RPE). This level of protection could also be obtained by wearing an integral suit with an external air supply. These suits are equipped with exhalation valves to expel excess air and maintain a positive pressure within the suit. They therefore give protection against low levels of fume, volatile liquids and particulates.

Type 1

When working in areas containing high levels of gases or volatile liquids, or when dealing with toxic chemicals (including potent dusty materials), a completely gas-tight suit should be worn. These can be worn either over breathing apparatus (Type 1(a)) or with the breathing apparatus strapped on the outside (Type 1(b)). Some gas-tight suits have air-lines to supply breathable air (Type 1(c)). The level of protection given by these suits will depend on the fabric of the suit. This is discussed in detail in the next section.

Fabric used for PPE

As explained in Chapter 7, chemicals can permeate through materials or degrade fabrics and so weaken the protection they provide. It is therefore important to give careful consideration to the fabric of the suit as well as to which type of suit to use.

In general, PVC will give adequate protection from all the common acids and alkalis but not the volatile organic chemicals. Neoprene, nitrile, butyl or hypalon offer better protection against these. These polymers are used as coatings on various fabrics such as polyester or nylon and different polymers can be combined to give an even better fabric. For example, a commonly used 'combination' fabric is hypalon layered on neoprene and butyl (HNB). This will give better protection than any of the single polymers. One of the best polymers is viton, a fluoro compound, which will protect against a wide range of organic chemicals, particularly when it is combined with butyl.

These traditional fabrics, which give adequate protection against a wide range of chemicals and are comfortable to wear, do have one disadvantage. If they are in contact with a particularly aggressive organic chemical for some time, that chemical will permeate into the fabric. It will permeate out again only slowly and if a suit has to be used again within a short period of time it may still be contaminated. Surface washing will remove contamination only from the surface, although it will speed up the rate at which the chemical leaves the fabric of the suit.

To overcome this problem, new polymers are being developed which will not allow organic chemicals to permeate. These new polymers are more polar than the traditional ones and offer a better barrier to the non-polar organic chemicals. Their main disadvantage is that they are soluble in water and/or acids and therefore have to be enclosed in protective layers of polyethylene. These are the new laminate fabrics, which will offer at least 8 hours' protection against aggressive organic chemicals. However, they are not as comfortable to wear because the laminates tend to be fairly rigid. They are classed as disposable because continual wear will crease the fabric, break the barriers and destroy the ability of the fabrics to act as barriers.

When deciding on which fabric to use, it is important to know how long the suit will be in contact with the chemical. This contact time should be less than the breakthrough time of the chemical and fabric concerned. If the risk concerned is one of occasional splash contact, then material with a shorter breakthrough time can be used, as a small splash will evaporate from a suit before the liquid has time to permeate, making the actual breakthrough time effectively longer.

A very important factor to consider is the hazardous nature of the chemical to be encountered. While some chemicals have obvious immediate effects, such as the corrosive effect on the skin of various acids, others have insidious longer-term effects. For example, aniline will slowly permeate into the skin and cause physiological damage. Other chemicals that invade the body by skin contact or inhalation include known carcinogens such as benzene or chloroform. With highly toxic chemicals it is important to choose the maximum protection. Fabrics should be chosen that have the longest breakthrough times and the lowest permeation rates.

Some typical examples of PPE that may be needed for maintenance work on containment systems are listed in Table 9.4. These are for guidance only: actual or proposed operations must be evaluated individually. To make the correct choice of PPE all of the factors described above should be considered and discussed with the company occupational hygienist and safety expert. Help can also be obtained from the manufacturer of the PPE, who will be able to supply permeation and degradation data.

The requirement for PPE usage during cleaning of containment devices depends on the actual or foreseeable exposure to the hazardous material.

For example, changing the local filter in a Containment Strategy 2 device may expose the maintenance fitter to more dust than the operator would be exposed to under normal use. Also, opening up isolators, even after internal cleaning, may still have the potential to release material. Removal of safe-change HEPAs may also expose maintenance fitters to hazardous concentrations and so PPE is deemed a necessary precaution.

Table 9.4 Typical PPE requirements for maintenance

Containment Strategy 2	Containment Strategy 3	Containment Strategy 4
Gloves Filtered air-fed safety helmet	Pharmaceutical gowning — disposable suit Over-shoes Gloves Filtered air-fed safety helmet Air-fed half-suit Full air-fed safety suit	Disposable under-suit Air-fed safety suit Inner gloves

151

Selection process

The steps to be carried out before buying any protective clothing are as follows:

1. Decide on type of clothing — Type 1(a), 1(b), 1(c), 2, 3 or 4;
2. Decide on fabric using the following criteria:
   - which chemical(s) will be encountered?
   - how long will the clothing be in contact with the chemical(s)?
   - what is the hazardous nature of the chemical(s)?
   - will the chemical(s) degrade the fabric?
   - if so, how rapidly?
3. Make sure all the experts have been consulted:
   - company health and safety personnel;
   - manufacturers of PPE;
   - maintenance technicians.

Standard operating procedures (SOPs)

It is essential to operate and maintain containment devices in a disciplined way. Incorrect operation may cause a breach of containment with dire effects on the safety and health of employees. For this reason, Standard Operating Procedures (SOPs) must be clearly written down and rigorously applied. These SOPs define a logical series of duties for the process operators, supervisors, managers, visitors, maintenance personnel and any others who may be involved, which, if carried out faithfully, ensure that the processes and equipment are operated safely and in accordance with current guidelines on good manufacturing practice (cGMP) and containment practices.

A typical SOP includes the following key features:

- outline of purpose and scope of the procedure;
- instructions for executing the procedure correctly;
- responsibilities and reporting structure for operators and all others involved in implementing the procedure;
- references and appendices.

The following sections provide examples of SOP instructions, which can be used as a framework for specific SOPs for particular transfer or maintenance operations. The examples given cover:

- donning or removing protective clothing;
- dispensing and subdivision of solids;
- cleaning and wash-down of transfer devices, process equipment and PPE.

Some of these example procedures may be applicable to operations, others to maintenance.

> However, it is important to remember that employers are responsible for devising procedures appropriate to their sites and operations. The examples below cannot be taken as correct for any similar application unless they have been expanded to include the required specific details exactly appropriate to that application. IChemE and the authors of this guide cannot, therefore, accept any responsibility for any consequences resulting from following these procedures as they stand.

In all cases, these procedures:

- assume that:
  — the hazard rating for the application and operations concerned have been defined;
  — a Containment Strategy has been selected;
  — appropriate containment devices, cleaning materials and PPE (e.g., as shown in Table 9.4) have been provided;
  — workers are not assigned to tasks where the materials being used may cause them allergic or physiological reactions;
- apply to Containment Strategies 2–4 only; for Containment Strategy 5 special procedures need to be devised.

Example SOP 1: Changing procedure

This SOP will be used to ensure that maintenance personnel have adequate protection when carrying out maintenance, including routine cleaning down and emergency repairs.

Its objective is to ensure that personnel are suitably attired in the correct protective clothing appropriate to the hazard bands of the hazardous materials used in it and the operations to be carried out that may expose employees to toxic materials.

This SOP comprises two separate procedures, for:

- entering the hazardous area;
- leaving the hazardous area.

The steps to be followed in each case depend on the Containment Strategy adopted for the area and are detailed in Table 9.5 (page 155). These are to be followed whenever anyone enters or leaves an area to which the SOP applies.

A similar procedure would be needed for donning and removing any other types of PPE required during routine operations under Containment Strategy 2. The procedure described here is not appropriate as it stands but could be suitably adapted for such a purpose.

Example SOP 2: Dispensing and subdivision of solids

This SOP is used by process operators to divide the contents of a bulk drum (i.e., raw materials or intermediate products) into batches ready for use in a process.

Its objective is to ensure that materials are weighed out to meet process batch weights in a manner appropriate to their hazard rating. The steps to be followed in each case depend on the Containment Strategy adopted for the material being handled and are detailed in Table 9.6 (page 156). On completion of this procedure, the area should be cleaned in accordance with SOP 3.

For some applications using Containment Strategy 2, appropriate PPE may need to be donned before applying this SOP and removed afterwards. A separate procedure, similar to SOP 1, should be prepared for this purpose.

Example SOP 3: Cleaning and wash-down procedure

This SOP is used by operators to clean the vessels, pipes and surfaces that may have become contaminated during the transfer. These operations should wherever possible be carried within the appropriate enclosure and under cover of the appropriate containment systems, so PPE should not be needed.

If more extensive cleaning down of contained areas is to be carried out, this should be done by maintenance engineers following a separate procedure. If there is any reason to believe there may be a risk of personal contamination, then before and after following that procedure, the engineers should respectively don and remove any PPE required in accordance with SOP 1. PPE may also be needed for routine cleaning of some Containment Strategy 2 applications, in which case a similar procedure for donning and removing it should be followed,

The objective of this SOP is to ensure that workplaces are cleaned in a manner appropriate to the hazard rating of the materials being handled.

The steps to be followed in each case depend on the Containment Strategy adopted for the area and are detailed in Table 9.7 (page 157).

Conclusion

This chapter has summarized the requirements to be met when training operators to use containment equipment and when maintaining equipment. The latter task may require the use of PPE, which has also been discussed here. The final chapter of this guide discusses the validation of containment equipment, to verify that it is suitable for its purpose and maintained in good working order.

Table 9.5 Changing procedure before and after maintenance

	Containment Strategy 2	Containment Strategy 3	Containment Strategy 4
Entering the hazardous area	• gain permission from plant supervisor; • sign log book; • enter change area; • don protective clothing; • exit to plant.	• gain permission from plant supervisor (including confirmation that materials being used will not cause allergic or physiological reaction); • sign log book; • store street clothing; • don protective clothing; • clothing check by plant escort; • exit to plant via step-over or equivalent entry point.	• gain permission from plant supervisor (including confirmation that materials being used will not cause allergic or physiological reaction); • sign log book; • store street clothing; • don disposable under-suit; • enter over-suit change area; • don air-fed safety over-suit; • check safe operation (by plant escort); • enter contained area via specific route; • ensure plant escort is in attendance;
Leaving the hazardous area	• enter changing area; • remove contaminated clothing in designated area; • wash hands; • consider photocopying of any paperwork/notes; • exit from changing area; • sign out and notify plant supervisor of departure.	• enter changing area; • consider decontamination of any tools or instruments used (e.g., tape measure); • remove contaminated clothing in designated area; • wash hands; • replace street clothing; • consider photocopying or faxing of any paperwork, notes, etc.; • exit from changing area; • sign out and notify plant supervisor of departure.	• enter over-suit decontamination area with plant escort; • photocopy/fax paperwork, notes, etc., from contaminated zone and store originals in the zone or dump in designated disposal bin; • fully wash down and de-contaminate air-fed safety over-suit; • dry over-suit and check for any contamination remaining; • repeat wash-down, drying and checking if necessary; • enter over-suit storage area; • remove over-suit and store; • enter shower area; • remove disposable under-suit and dump in designated disposal bin; • shower; • replace usual street clothes; • exit from changing area; • sign out and notify plant supervisor of departure.

Table 9.6 Dispensing and subdivision of solids

Containment Strategy 2	Containment Strategy 3	Containment Strategy 4

Containment Strategy 2	Containment Strategy 3	Containment Strategy 4
• position materials into weighing area; • earth-bond materials; • select suitable scoops or hand tools; • dispense only to elbow depth; • ensure posture upright; do not lean over drum; • beyond elbow depth use long-handled scoop and incline drum.	• position materials into handling chamber; • earth-bond materials; • select suitable scoops or hand tools; • inspect, position and dock any receiving containers, drums, etc., to the isolator outlet port; • check all weighing systems are operating to specification; • where an antechamber is provided for removing drum lid, ensure drum's inner liner is sealed and no loose powder is visible before proceeding to dispensing stage. Where liner is damaged or open, follow emergency de-contamination procedure; • offer/raise drum to powder-handling chamber interface. Open drum only when all seals/ports are in the correct position; • carry out subdivision as appropriate and load materials into process vessel; • seal drum and de-contaminate it by wiping it down in the handling chamber, using glove-ports; • remove drum from handling chamber as soon as practicable, ensuring any transfer ports are closed; • prepare all scoops, weigh platforms, etc., for de-contamination; • observe all special cleaning procedures; • remove all scoops, etc., to wash bay via protective polythene bags.	• position materials in airlock or adjacent to Rapid Transfer Port (RTP) or flexible liner tube; • transfer materials to isolator via appropriate transfer device; • make direct connection between supply and receiving container by approved method; • initiate transfer of material through this direct connection to desired weight; • clean down interior of contained area if required; • remove any empty containers and/or weighed batches via airlock or RTP or flexible liner tube; • carry out final de-contamination of airlock, RTP flanges, etc.

Table 9.7 Cleaning and wash-down

Containment Strategy 2	Containment Strategy 3	Containment Strategy 4
• remove all bulk containers and drums from area; • gain authorization from plant/area supervisor to commence clean-down; • vacuum up gross powder spillages to minimize loadings on plant effluent treatment plant; • hose down general area using brush or sponge for detailed cleaning; • dry down main surfaces; • gain approval from supervisor that cleaning is satisfactory.	• remove all containers and drums from device using transfer port or other appropriate method; • gain authorization from plant/area supervisor to commence clean-down; • wash down all internal major surfaces using a wash lance or CIP system (with approved cleaning agents); • dry down main surfaces, using hot-air dehumidification if required; • clean main sinks and drains with approved disinfecting solution or other approved cleaner as required; • test gloves for damage and porosity; • gain approval from supervisor that cleaning is satisfactory.	• gain permission from area supervisor to commence clean-down procedure; • for automated CIP systems commence cycle; • for manual wash-down systems use approved cleaning hose or other washing system; • pay attention to corners and glove interfaces; • remove contaminated water to drain/effluent system; • effect drying cycle or manual wipe-down; • inspect half-suit for pinholes and other damage; • test gloves for damage and porosity; • gain approval from supervisor that cleaning is satisfactory.

Note: For some glove-box isolator applications involving volatile or aseptic materials, an inert or sterile environment may be required. For this reason, re-testing will be necessary after cleaning procedures have been completed, to verify that conditions are safe.

157

Validation

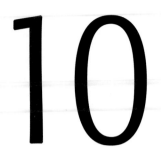

Purpose

The purpose of this chapter is to define the processes by which systems, both hardware and software, can be subjected to tests that provide documented evidence that they meet their required performance criteria. The approach defined in this chapter applies mainly to pharmaceutical processes but the principles can be applied to other industries as well.

Contents

What is validation?

Validation is the confirmation by examination and provision of objective evidence that the particular requirement for a specific intended use can be consistently fulfilled.

In the case of containment systems, validation includes a demonstration that hazardous materials are adequately contained and hence that concentrations of those materials in areas where people are likely to be present are at acceptable levels. However, it does not end with acceptance of the system into operation; monitoring should be carried out continually to demonstrate that the system remains effective and to enable prompt corrective action to be taken should levels of contaminant in the workplace reach unacceptable levels.

The principles of validation described in this chapter have been developed in the pharmaceutical industry but can be usefully applied to the design, selection and application of containment equipment in any industry.

Although this guide is concerned specifically with containment systems, the principles described in this chapter can be applied to the whole plant as well as to its specific provisions for containing hazardous substances.

Benefits of validation

Compared with processes not subjected to the same degree of inspection, a properly validated and controlled process will in general:

- yield fewer products that are out of specification, produce increased output;
- achieve enhanced performance;
- result in significantly lower exposure of employees to the materials to be contained.

The data collected in validation files can prove invaluable during investigations of inadequate performance of containment equipment and in developing improvements in the design or specification of the next generation of containment equipment.

Guide to the validation process

An essential requirement of the validation process is that each stage of the specification, design, construction, installation, testing and maintenance of a containment system should be clearly documented. Properly certified documentation is the only evidence acceptable to pharmaceutical inspection authorities as proof that work has been done and approved to an agreed standard.

The key stages in the validation process are:

- preparation of a **Validation Master Plan** (VMP);
- *Design Qualification* (DQ): verifying that the design documentation correctly specifies the required system;
- *Installation Qualification* (IQ): verifying that the system is built and installed in accordance with the design documentation;
- *Operational Qualification* (OQ): verifying that the system functions as required;
- *Performance Qualification* (PQ): verifying that the outputs from the system (e.g., products or intermediates) conform to requirements;
- *Computer systems validation*: verifying that software components of the system function as required.

These stages are summarized in the remainder of this chapter and Figure 10.1 (page 160) gives an overview of the validation process as it could be applied to a contained process system. It should be noted that Figure 10.1 and subsequent figures referenced from it are concerned only with validation actions; in practice, other activities will take place between the stages shown. For example, equipment will generally be supplied and installed between the

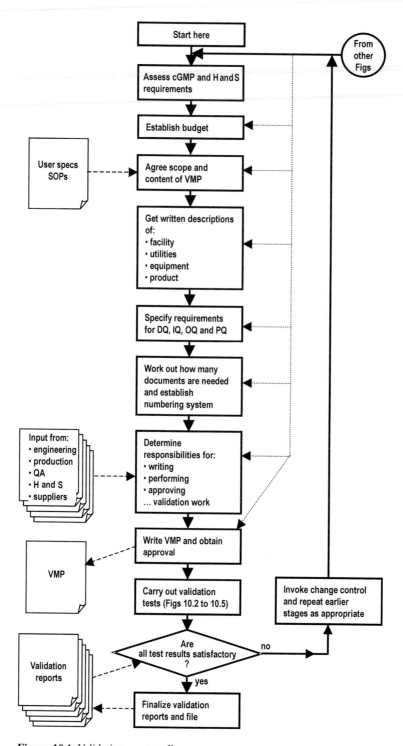

Figure 10.1 Validation — overall process

DQ and IQ stages. More detailed information on the technical criteria that containment systems must satisfy can be found in other chapters in this guide. More details about the validation process can be found in guidance produced by the Medicines Control Agency (1997), known colloquially as the 'Orange guide' and in the US, Code of Federal Regulations.

Who should be involved?

To ensure that all validation issues are identified and that appropriate steps are taken, all the following people should typically be involved during the course of the process (not necessarily at every stage):

- project manager;
- production representative;
- Quality Assurance (QA) representative;
- technical representative;
- maintenance representative;
- safety representative;
- occupational hygienist.

The group may be extended to include laboratory staff and specialist advisers as appropriate.

Validation Master Plan (VMP)

Careful planning of the validation process is essential to ensure that the whole issue of validation does not become an administrative burden rather than an essential design and operating tool. The VMP is a structured, detailed plan of work, which summarizes the design basis for a given project, defines the scope of all the required validation tasks (DQ, IQ, etc.) and describes how those tasks are to be carried out and controlled. The VMP is not a static document — it is continually updated as information becomes available.

In essence the VMP answers the question 'how will you know when the validation work for this project is complete?'

Typical elements in a VMP are:

- identification of process, system or equipment to be validated;
- criteria for success;
- duration of the process;
- assumptions;
- identification of utilities used and definition of their quality;
- identification of operator qualifications and training requirements;
- description of the process and equipment;
- relevant specifications for the product, components, manufacturing materials, the environment, etc.;
- any special controls or conditions to be used during validation testing;

161

- product or performance characteristics to be monitored and methods to be used;
- any subjective criteria;
- definition of what constitutes 'non-conformance' for both measurable and subjective criteria;
- statistical methods for data collection and analysis;
- maintenance and repair requirements;
- stages in the process at which review is required;
- conditions that would necessitate re-validation;
- approval of protocols.

Change control

Any modification made to a process could have unforeseen and possibly serious consequences for the operation and safety of the process. It is, therefore, essential to have a formal change control procedure in place to ensure that the implications of any proposed change are fully assessed and documented before the change is implemented.

A change to the equipment could be required to rectify a failure to implement the design properly. It could also constitute a change to the design itself. For this reason the change control procedure must allow for the possibility that a deficiency detected during the IQ, OQ or even PQ stage may result in the need to repeat the DQ process to evaluate the implication of the proposed corrective action and confirm the integrity of the design. Alternatively, the deficiency may identify a flaw in the testing protocol used and this would need to be revised.

In theory, therefore, a problem identified at any stage could require a return to the start of the whole validation process to reassess the criteria, scope or even the budget. In practice, some of the earlier stages may be omitted provided that the decision to do so can be justified and the reasons for doing so are documented.

The flow charts for the various stages allow for these considerations by including tests that could result in a return to an earlier stage of the validation process, whilst accepting that it may be a matter of judgement as to which step is actually followed next.

The importance of documentation in the validation process cannot be overstated. When equipment is relocated, replaced or maintained, good documentation can obviate part of the need for re-validation. Indeed, in the eyes of many pharmaceutical regulators, a machine without accurate validation and maintenance records does not officially exist.

Design Qualification (DQ)

Design Qualification (DQ) is the production of documented evidence that the design of equipment, facilities and operations complies with quality specifications. In the case of purpose-built plant, DQ will normally be performed on the design documentation before the equipment is built; for off-the-shelf equipment it is carried out before the purchase order is placed. Figure 10.2 details the DQ process.

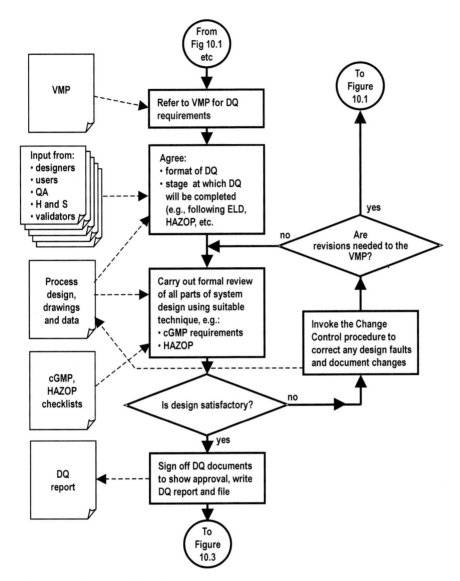

Figure 10.2 Design qualification

The definition above fulfils the requirements for DQ with regard to validation. However, DQ is also about getting the project right first time. The designers of a project should consider:

- health and safety, including:
  — the risks presented by materials handled;
  — the Containment Strategy required to control those risks;
- environmental issues;

163

- engineering concerns, e.g.:
  - scale-up from laboratory process to full production;
  - constructability;
  - maintainability;
- business concerns, e.g.:
  - cost-effectiveness.

DQ can be applied to processes as well as equipment. The more issues that can be identified at the design stage of the project, the easier it will be to provide a system that will be suitable for its intended use. Central to the success of the DQ process is a formal methodology, such as a Hazard and Operability (HAZOP) study for reviewing the Engineering Line Diagrams (ELDs), Piping and Instrumentation Diagrams (P and IDs), Operation and Maintenance (O and M) manuals or other design documentation. The HAZOP method is described by Kletz (1999) and by Tyler, Crawley and Preston (1999) among others.

DQ can be performed at two levels:

- *Level 1 DQ* is a high-level study and is performed as part of a project feasibility study to ensure that all quality aspects are identified and their impact on cost established before the project starts. The findings must be included in the *User Requirements Specification* (URS), *Functional Design Specification* (FDS) and *Design Specification* (DS).
- *Level 2 DQ* is a more detailed study, which focuses specifically on the quality aspects of the equipment and process. For example, the Level 1 study would identify a list of critical devices as well as temperature ranges, electrical standards, cleaning procedures, etc. that must be complied with to satisfy the required quality criteria. The Level 2 study would investigate these critical elements in more detail, including the expected performance of the system at the extremes of temperature or other critical parameters.

In the above definitions:

- a *critical device* is any control, measuring, monitoring or testing device whose failure may have an adverse effect upon product quality, product security or integrity of packaging ('failure' in this context includes complete or partial failure, intermittent operation, ineffective operation or calibration defects).
- a *critical stage* of a process is an operation during a production activity that, if not controlled within predetermined limits, may affect product quality.

Typical items for consideration in a Level 2 DQ include:

- people:
  - is the system user-friendly?
  - is specialist knowledge required?
  - will training be required?
  - if so, what level of training?
  - will personnel involved be based on or off the site?
  - can the supplier produce a tailor-made package?
  - what product-critical tasks do personnel carry out?

164

- process:
  — are there any regulatory requirements?
  — if so, what are they?
  — what are the product quality specifications?
  — will specialist safety or environmental services be required?
  — what are the critical stages and process parameters?
  — how will extremes of temperature, pressure, etc., affect the end product or its intermediates?
  — will there be sufficient space in the work area around the equipment to perform manufacturing and maintenance tests?

Installation Qualification (IQ)

After its design has been validated by means of the DQ procedure, process equipment can be installed. It should then be reviewed, calibrated, challenged and evaluated to verify that it conforms to its design documentation and is capable of operating within established limits and tolerances, as well as throughout all anticipated operating ranges.

Installation Qualification (IQ) is a documented demonstration that equipment is installed as designed and specified and is correctly connected to with its associated plant. The flow chart for the process is shown in Figure 10.3 (page 166).

The key to a successful IQ is a rigorous set of protocols, which must be carefully prepared to ensure that all required tests are carried out and documented. Points to be addressed when preparing IQ protocols should typically include:

- determining installation procedure;
- determining necessary environmental controls and procedures;
- determining requirements for calibration, cleaning, maintenance, adjustment and expected repairs;
- examining equipment;
- comparing it with supplied documentation (validated by DQ);
- verifying correct installation;
- carrying out initial calibration and ranging;
- reviewing, checking and distributing manuals and spares lists.

Prior planning for eventual maintenance and repairs can reduce or prevent confusion during emergency repairs, which could lead to improper repairs such as the use of the wrong replacement part.

Operational Qualification (OQ)

Once IQ has verified that the system has been installed correctly, *Operational Qualification* (OQ) is carried out. OQ is a documented demonstration that facilities and operations function as specified throughout the anticipated operating ranges. It is normally carried out using materials that simulate the actual process chemicals as closely as possible but with hazardous

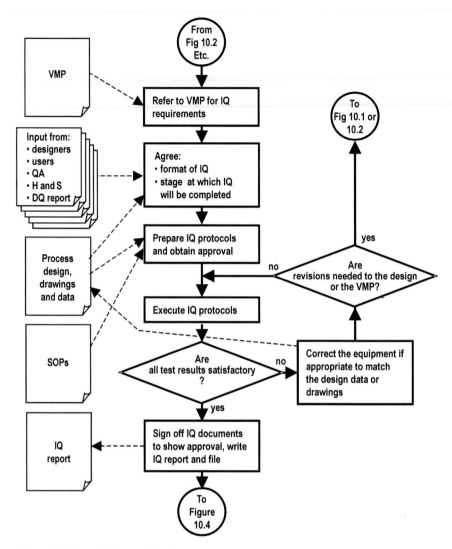

Figure 10.3 Installation qualification

substances replaced by harmless alternatives, or ***placebos***, to reduce the risk of exposure in the event of a leak, rupture or other incident.

These placebos must be measurable, so that their concentrations can be monitored inside and/or outside the containment equipment in order to demonstrate that the required level of containment is indeed achieved. The flow chart for the process is shown in Figure 10.4.

Other useful outcomes of OQ activities could include:

- preparation of production documents and SOPs relating to:
  — operation;

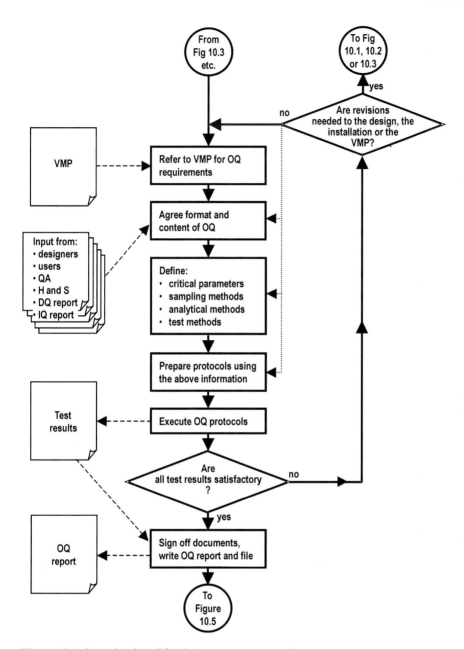

Figure 10.4 Operational qualification

  — in-process control;
  — cleaning;
  — data recording;
- preparation of maintenance documentation:
  — set-up and change-over procedures;
  — list of spares.

OQ protocols should be written to interrogate the operation of the containment equipment as a whole, not just individual parts. They should confirm that the equipment operates within its design specification over a range of speeds, batch sizes, air flows, etc. In the case of containment systems it is essential the protocols should cover checking that exposures of personnel to hazardous substances, or in this case the placebos replacing them, remains within acceptable limits. As a key purpose of the containment is to protect individuals, the most realistic test of its effectiveness is achieved by personal sampling as described in Chapter 3. However, it may also be prudent to sample exposures over time at various locations in the workplace. Similarly, if the purpose of the containment is to protect the product rather than the person, it should be possible to monitor levels of contaminant inside the equipment.

However, it should be borne in mind that challenges to the process posed during OQ tests should simulate conditions that will be encountered during actual production. The process should be made to operate at its allowed upper and lower limits of pressure, temperature, etc., in order to verify that adequate containment is maintained even under these extreme conditions.

Equipment manufacturers should perform qualification runs at their facilities and analyse the results to verify that the equipment is ready for delivery. Results from these qualification runs can be a valuable source of basic data and should be used as a guide to the overall performance of the equipment. However, it is usually insufficient to rely solely on the results obtained by the manufacturer and it should be borne in mind that responsibility for evaluating the equipment and challenging its performance rests with the company using it.

Process and monitoring equipment should be calibrated at the beginning of the IQ tests and the calibration checked at the end of the OQ tests to establish confidence in the validation of the process. If the monitoring equipment is found to be out of calibration at the end, it may indicate that the process has been out of control. In that case, the containment equipment and the process of which it forms a part cannot be considered to be validated and the protocols should be repeated, with amendments to include more frequent calibration checks.

Performance Qualification (PQ)

The purpose of Performance Qualification is to test the process to determine whether it is capable of consistently producing an output that meets specifications. It differs from OQ in that the materials used will be those actually required during production rather than inert or non-harmful substitutes. Although by this stage the OQ will have demonstrated that adequate levels of containment are achieved for placebos, it is necessary during PQ to demonstrate that they are also achieved for process chemicals. Monitoring of concentrations around the process equipment is therefore essential and those carrying out the tests should wear appropriate PPE (as discussed in Chapter 9) until it has been demonstrated that the concentrations of hazardous chemicals in the workplace are acceptable. Monitoring should also be carried out inside the process equipment if the product is to be protected from contamination. The flow chart for the process is shown in Figure 10.5 (page 170).

As in the case of OQ, the challenges to the process specified in the PQ protocols should simulate conditions that will be encountered during actual production. The protocols should also be designed to check that the process is *robust*, i.e., that the quality of its products remains consistent under all conditions, by forcing it to operate at its allowed upper and lower limits of pressure, temperature, etc. For this reason, it is necessary to identify the critical parameters in the process so that the PQ protocols can test their effects thoroughly.

Since PQ is concerned with product quality, it is especially important to involve QA personnel in its planning and execution, as they will need to have the resources ready to test the large quantities of sample product that will be sent to them for testing.

Validation of automated systems

The introduction of computers and automated manufacturing systems has led to an increased focus on such systems and their regulatory compliance. Important stages in the development of an automated system include production of the documentation shown in Table 10.1 (page 171).

The validation required includes detailed reviews of this documentation against the source code to verify that the software actually achieves the tasks required of it. Details of how this and other validation tasks for automated systems should be performed can be found in ISPE (2001). The link between these documents and the validation activities discussed earlier in this chapter is shown in Figure 10.6 (page 171), adapted from ISPE (2001).

Validation review

Before the validation can be reviewed, the constituent elements, i.e., DQ, IQ, OQ, and PQ, must be completed. If all the work has been progressed through all of the stages described in this chapter, all the necessary documents will now be in place. The following completed and authorized documents must be available for review:

- DQ report and associated specifications;
- IQ report;
- OQ report;
- computer validation rationale;
- computer test protocols, with results;
- source code review;
- PQ report.

All of these documents will be reviewed for compliance with validation requirements; the validation status of each stage will be recorded in the validation review report. All actions outstanding must be identified and dates for their completion recorded. The *validation review report* makes a final statement of acceptance or rejection of the specified process.

169

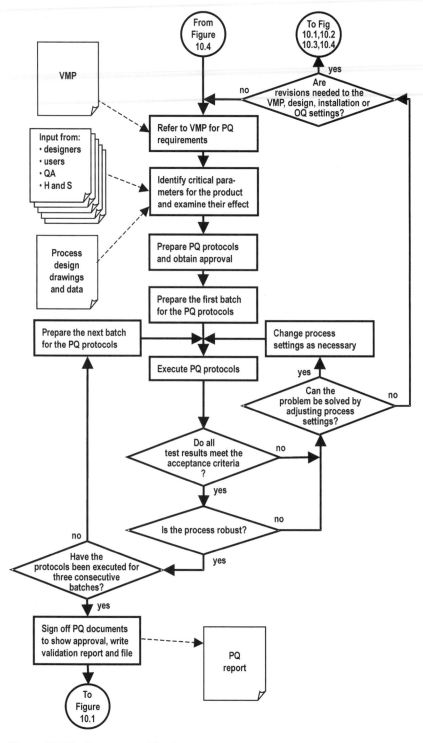

Figure 10.5 Performance qualification

Table 10.1 Development of an automated system

User Requirement Specification (URS)	The URS is produced by the user to define clearly and precisely what the system is expected to do and to state any constraints, regulatory and documentation requirements.
Functional Design Specification (FDS)	The FDS defines the function that a system or part of a system should perform. It includes consideration of *how* and *by which elements* a function has to be implemented. It is normally written by the supplier of the equipment.
Design Specification (DS)	The DS is a complete definition of the equipment or system in sufficient detail to enable it to be built.

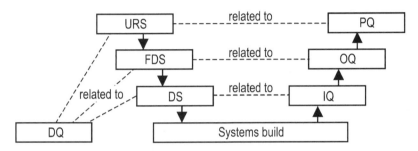

Figure 10.6 Development stages and validation activities

Conclusion

This chapter has summarized the procedures required to ensure that a process, including any containment systems it contains, performs in accordance with the design criteria for their safe operation. The requirements for containment within a process will have been identified and appropriate equipment selected and installed correctly and sound procedures for operating them drawn up in accordance with the principles described in earlier chapters of this guide. After this, the installation and performance of the containment systems must be validated by implementing the procedures described in this chapter. When all these stages have been completed and programmes established for ensuring that the equipment is maintained in good working order, the process can safely be put into operation.

Appendix 1 – Glossary

This glossary lists terms and abbreviations and defines how they are to be interpreted in the context of this guide. Terms presented in *italics* in the definitions are themselves defined elsewhere in the glossary.

Acceptable ceiling concentration
The maximum airborne concentration of a *substance* to which employees may be exposed at any time except for designated time periods during which the *TWA* concentration to which they are exposed must remain below the designated *acceptable maximum peak* level for that *substance*.

Acceptable maximum peak
The maximum airborne concentration of a *substance* to which employees may be exposed during excursions above an *acceptable ceiling concentration*.

Acute effects
Effects on the human body that occur a short time after exposure to a *substance hazardous to health*.

Active sampling
A method of measuring the concentration of a gaseous contaminant in the atmosphere by drawing air through a filter medium onto which the contaminant will adsorb.

ACOP
Approved Code of Practice, a guidance publication based on *Regulations* which, if followed, helps to achieve compliance with the law (UK).

Active valve
The part of an *SBV* that seals the inlet connection of a vessel designed to receive material by direct transfer through the *SBV*.

Administrative control
Any procedure that significantly limits daily personal exposures to chemical, physical or biological agents by controlling or manipulating the work schedule or the manner in which the work is performed.

AGV
Automated guided Vehicle, a robot truck for transporting materials.

Alpha-beta port
Another name for *RTP*.

Air-suit
A type of *PPE* comprising a complete suit with integral visor and breathable air supplied either from cylinders carried on the operator's

back or by means of a flexible pipe linked to a remote source of clean air.

Auto-ignition temperature The lowest temperature at which a mixture of vapours, gases and/or dusts in air will ignite spontaneously.

Batch splitting Weighing out discrete quantities off-line of solids or liquids required for a process batch, either raw materials bought into the plant or intermediates made by previous process steps.

BATNEEC (1) In EU Directives, Best Available Technology Not Entailing Excessive Cost.
(2) In the *EPA*, Best Available Technique Not Entailing Excessive Cost, where the term 'technique' covers both technology and operational factors.

Breakthrough time The time taken for a given *chemical* to pass through a given barrier.

Butterfly valve A disc the diameter of the pipe in which it is located and capable of rotating 90° about an axis through its diameter so that it either permits fluids or powders to flow past it or blocks the flow.

Capture velocity The minimum air velocity required to capture or divert by bulk air entrainment a cloud of dust or vapour into an exhaust *hood*.

Carcinogen A substance liable to cause cancer if ingested, inhaled or otherwise taken into the body.

Category of danger A description of *hazard* type as defined in *CHIP*, e.g. 'harmful', 'irritant', 'sensitizer', 'toxic'.

Ceiling level The maximum airborne concentration of a *substance* to which employees may be exposed at any time.

cGMP Current perception of Good Manufacturing Practice as defined in US Federal Regulations and equivalent EU Directives.

Chemical A common term for *substances* and *preparations*.

CHIP Regulations The Chemicals (Hazard Information and Packaging for Supply) Regulations 1994 as amended (UK).

Chronic effects Effects on the human body that become apparent over a long period of time after exposure to a *substance hazardous to health*.

CIP Cleaning in Place, arrangements made (usually automatically) to decontaminate containers, process equipment and the inside of *isolators*, without dismantling the equipment concerned, achieved by using liquid spray impingement to remove accumulated solids and/or liquids from their surfaces.

Classify
Identify the *hazard* of a chemical by assigning a *category of danger* and a *risk phrase* using set criteria.

Cleanability
One of four indications of the ease with which a piece of equipment can be cleaned: C1 = manual wash-down; C2 = dismantle and wash-down with a solvent wipe; C3 = decontaminated using *CIP* procedures; C4 = *CIP* with contained dismantling.

Closed transfer
The movement of materials between two containers or between a container and process equipment through a closed system of pipes or ducts or by direct connection.

Computer systems validation
Validation of computer systems or other programmable devices used for process control and/or data management.

Containment
(1) Generally throughout this guide, the enclosure of a process and/or control of the environment in which it is carried out in order to prevent the contamination of people by the materials used or vice versa.
(2) In the context of explosion control, designing a process to withstand the full explosive pressure that could occur in the event of an explosion.
(3) In HSG193, enclosure of a *hazard* with only very limited breaches of the enclosure permitted.

Containment Strategy
One of five generalized approaches proposed in this guide to limiting or eliminating exposures of personnel, products or the environment to hazardous concentrations of *substances*. Each provides protection at a specific level ranging from reliance on *general ventilation* for the removal of *substances hazardous to health* (Containment Strategy 1) to total enclosure and mechanization of the process with all contact between the operator and *substances hazardous to health* eliminated (Containment Strategy 5). Analogous to *Control Approach* but with more levels.

Control Approach
One of four generalized approaches to exposure limiting/elimination defined by the HSE in HSG193, comprising reliance on *general ventilation* for the removal of *substances hazardous to health* (Control Approach 1), *engineering control* (Control Approach 2), *containment* (3) (Control Approach 3) and reliance on specialist expert advice (Control Approach 4).

Control factor A measure of the effectiveness of a control, being the ratio of the exposure obtained without the control to that obtained with the control operating.

COSHH Regulations The Control of Substances Hazardous to Health Regulations 1999 (UK).

Critical device Any control, measuring, monitoring or testing device whose failure, including complete or partial failure, intermittent operation, ineffective operation or calibration defects, may have an adverse effect upon product *quality (2)*, product security or the integrity of the complete process.

Critical stage An operation during a production activity that, if not controlled within predetermined limits, may affect product *quality (2)*.

Cytotoxic Harmful to cells.

Data processing systems Computer systems, such as office personal computers or mainframe computers and including the software installed, used to aid business processes.

Declaration of conformity A *QA (1)* document that specifies the criteria to which a product was made and certifies that the product as supplied conforms to those criteria.

Degrading Loss of expected properties; deterioration of material, by corrosion, solvation or other changes.

Diffusive sampling A method of measuring the concentration of a gaseous contaminant in the atmosphere by allowing the contaminant to adsorb naturally onto a filter medium.

Dilution ventilation An alternative name for *general ventilation*.

Dispensing Another term for *batch splitting*.

Down-flow booth An enclosed area of a workplace into which air is supplied via the ceiling and from which it is extracted through a vent low down in the booth.

DPTE® Double Porte de Transfer Entaché, a type of *RTP*. DPTE® is a registered trade mark of La Calhène, Inc., Rush City, MN.

DQ Design Qualification, the preparation, as part of the *validation* process, of documented evidence that the design of equipment, facilities and operations complies with *quality (1)* requirements. Carried out in two stages: *Level 1 DQ* and *Level 2 DQ*.

Drum wand　A tube through which liquids are sucked from a drum or solids are transferred (usually by air suction) out of a container.

DS　Design Specification, a complete definition of the equipment or system in sufficient detail to enable it to be designed in detail and then built once the design has been approved.

Due diligence　A defence against legal action under health and safety legislation whereby the defendant demonstrates that (s)he had taken all reasonable precautions to avoid committing the offence.

Engineering control　The use of hardware, other than *PPE*, to control the physical environment in a manner that limits the exposure of individuals to the harmful effects of a particular *hazard*.

EP　Exposure Potential, a measure of the *risk* arising from the use of a substance in a process, taking into account the *toxicity* and physical characteristics of the *substance*, the quantities used and duration of transfer of the *substance* into or out of the process.

EPA　The Environmental Protection Act 1990 (UK).

Explosion suppression　Introducing *substances* that will suppress an explosion as soon as it occurs.

Explosion venting　The provision of pressure-relief panels on equipment or buildings that, in the event of an internal explosion, will blow out and so prevent the development of damaging excess pressures.

Exposure limit　A term used in this guide to cover *OES*, *MELs* and *PELs*.

FDS　Functional Design Specification, a document detailing the functions to be carried out by a computer system and how they are achieved.

Flammable　Liable to burn or explode if ignited.

Flash point　The lowest temperature at which a liquid gives off sufficient vapour in sufficient concentration to form a combustible mixture with air near its surface.

Full containment　*Containment (2)*.

Gas chromatography　A technique used during analysis of air samples to separate the various contaminants within a sample. An inert gas blows the sample through a column of adsorbent material. Molecules of different contaminants travel through the column at different rates and this enables their relative quantities in the sample to be measured.

General ventilation	Provision of an air supply or the reliance on natural air-flows to reduce the concentration of a contaminant released into the workplace.
Glove-box	An *isolator* fitted with *glove-ports*.
Glove-port	A non-porous glove with long sleeves (generally elbow length) sealed into an opening in the wall of an *isolator* so as to maintain the *containment (1)* provided by the *isolator* whilst allowing the operator to manipulate objects and controls inside it.
HAC	Hazardous Area Classification, dividing the workplace into 'zones' as defined in BS EN 60079-10 according to the explosion *risks* generated by the *substances* present and ensuring that all electrical equipment is approved for use in the zone where it is required.
Hazard	In general, anything with the potential to cause harm to people or to the environment. In the context of this guide it normally means the potential harm that can arise from use of a *substance*, reflecting the inherently dangerous properties of that *substance*.
Hazard group	One of six categories (A to F as proposed in this guide) to which a *substance* can be allocated depending on the severity of the *hazard* it presents to those exposed to it by inhalation.
Hazardous chemical	Any *chemical* that is a *physical hazard* or a *health hazard.*
HAZOP	Hazard and Operability study, a technique for performing a comprehensive analysis of a process to identify all the possible *hazards* it contains.
Health hazard	A *chemical* for which there is statistically significant evidence based on at least one study conducted in accordance with established scientific principles that acute or chronic health effects may occur in exposed employees.
HEPA	High-Efficiency Particulate Arrestor, a type of filter used to remove very small particulate contaminants from an air-flow.
Hierarchy of controls	A priority scheme for the selection of a control method for a *hazard* whereby elimination of the *hazard* is the preferred option, *engineering controls* and then *administrative controls* are applied if previous methods have failed to achieve adequate control and *PPE* is used only to control residual *risk* when all other methods have been applied and have reduced the *risk* to the greatest extent reasonably practicable.
Highly controlled area	An area to which no personnel have access and where complete emergency action plans and spillage procedures exist.

Hood

The aperture through which an *LEV* system draws in contaminated air to be removed from a working area.

Horizontal laminar flow booth

An enclosed area of a workplace from one closed end of which air is exhausted through a perforated distribution plate, which ensures that an even flow of air is drawn into the area from the open end.

HSC

Health and Safety Commision, established in the UK by the *HSW Act* to undertake research and propose *Regulations* on health and safety matters.

HSE

Health and Safety Executive, established in the UK by the *HSW Act* to assist the HSC in meeting its objectives. Its duties include publishing *Regulations*, with associated *ACOPs* and guidance material, on health and safety matters and enforcing the provisions of the *HSW Act* and *Regulations* made under it.

HSW Act

The Health and Safety at Work etc. Act 1974.

Inerting

The replacement of air inside process equipment with an inert gas (usually nitrogen) that will not affect the process but by excluding oxygen will prevent an explosive mixture from forming.

Inhalable dust

Airborne material that is capable of entering the nose and mouth during breathing and is thereby available for deposition anywhere in the respiratory system.

Intrinsic safety (electrical)

The property of an electronic or electrical device and its circuit that ensures that the amount of energy released at the device in the event of a failure is below that required to form a spark.

IQ

Installation Qualification, a documented demonstration, as part of the *validation* process, that facilities and equipment are installed as designed and specified.

Isolator

A sealed enclosure within which operations can be carried out without exposing the operators or the surrounding environment to contamination from the process materials inside it or vice versa.

LEL

Lower Explosive Limit, the concentration of gas or vapour in air below which the gas atmosphere cannot be ignited.

LEV

Local Exhaust Ventilation, a system for drawing air in the immediate vicinity of a source of contamination away from those working in the area.

Level 1 DQ

A high-level study performed as part of a project feasibility study to ensure that all *quality (2)* aspects are identified for cost analysis and incorporated in the *URS* or *DS* before the project starts.

Level 2 DQ	A more detailed study than *Level 1 DQ*, performed on the completed *URS* and *DS* of the system and focusing specifically on the *quality (2)* aspects of the equipment and process.
LFL	Lower Flammable Limit, the concentration of gas or vapour in air below which the gas atmosphere is not *flammable*. In practice, identical to the *LEL*.
Liquid chromatography	A technique similar to *gas chromatography*, in which a liquid solvent is used in place of the inert gas to transport the sample through the adsorbent column.
Limiting oxygen concentration	The percentage of oxygen in an atmosphere above which the atmosphere becomes explosive.
Limited venting	Exhausting excess gases into a confined area in the event of an explosion.
Long-term exposure	The airborne concentration of a *chemical* in a person's breathing zone, calculated as a *TWA* over 8 hours.
Machine	A piece of equipment that has moving parts and, usually, some kind of drive unit.
Machinery Directive	A colloquial name for the European Council Directive 89/392/EEC, subsequently amended by various Directives consolidated under 98/37/EC. Also used colloquially in the UK to mean the Regulations derived from it, namely the Supply of Machinery (Safety) Regulations 1992 and the Provision and Use of Work Equipment Regulations 1992.
MEL	Maximum Exposure Limit, the maximum permissible concentration of a *chemical* to which personnel may be exposed measured as a *TWA* over a stated reference period (15 minutes for short-term limits; 8 hours for long- term limits); set for *substances* that may cause serious health effects and for which either no safe levels exist or control to safe levels is not reasonably practicable. Exposures to such *substances* should be reduced so far as is reasonably practicable but must never be allowed to attain or exceed the MEL.
MSDS	Material Safety Data Sheet, information that must be provided as part of the *supply requirements* under the *CHIP Regulations* in the UK or 29 CFR 1910.1200 in the US.
Mutagen	A *substance* liable to change genes if ingested, inhaled or otherwise taken into the body.
Non-conformance	Failure of a process, procedure or application thereof or resulting product to comply with one or more specific *quality (1)* criteria.

OEL Occupational Exposure Limit, a term covering both *OESs* and *MELs*.

OES Occupational Exposure Standard, a standard concentration of a *chemical* at which there is currently no evidence that personnel are likely to be harmed as a result of exposure to that *chemical* at that concentration calculated as a *TWA* over a stated reference period (15 minutes daily for short-term limits; 8 hours for long-term limits) are likely to be harmed as a result. Exposures to a *substance* should not exceed its OES.

Open transfer The movement of materials between containers by scooping or pouring from one container to the other.

OQ Operational Qualification, a documented demonstration, as part of the *validation* process, that facilities and operations function as specified throughout the anticipated operating stages.

Parenteral Administered by some means other than through the alimentary canal; normally by injection through the skin.

Passive valve The part of an *SBV* that closes off and seals the bottom of a container.

PEL Permissible Exposure Level, the maximum airborne concentration of a *chemical* to which personnel may be exposed, which takes the form of a *TWA* exposure, *ceiling level* or *acceptable ceiling concentration* with *acceptable maximum peak* value.

Penetration The gradual transfer of material through holes in an apparently solid barrier.

Permeation The gradual transfer of molecules of a liquid or gas by diffusion between the molecules of a solid barrier.

Permeation rate The rate at which a *chemical* passes through a given sample of a material.

Physical hazard A chemical for which there is scientifically valid evidence that it is a combustible liquid, a compressed gas, explosive, flammable, an organic peroxide, an oxidizer, pyrophoric, nonstable (reactive) or water-reactive.

Placebo A harmless *chemical* substituted for a *substance hazardous to health* during *OQ* to enable the performance of a process to be simulated without the danger of exposing people to that harmful substance in the event of a leak, rupture or other incident.

Potent (Of a *substance*) able to produce significant physiological effects on the body even if present in very small quantities.

PPE	Personal Protective Equipment, any equipment worn by individuals to protect them against workplace *hazards*. In the context of general workplace safety it can include such items as hard hats, safety harnesses, ear-muffs or reinforced boots. In the context of this guide it refers to garments, from gloves to *air-suits*, that protect the wearer against harmful *substances*. PPE is to be used only where it is not reasonably practicable to control the *risk* adequately by other means.
PQ	Performance Qualification, a programme of documented testing, as part of the *validation* process, to show that a system (or group of systems), when performing with process material within product-specific design parameters, will consistently meet predetermined acceptance criteria.
Preparation	A mixture of *substances*.
Pressure-shock resistance	The ability of an item of equipment or a part of a building to retain its structural integrity under the impact of a rapid pressure rise caused by a dust or vapour explosion.
Production control systems	Computer systems, including all of the associated hardware and software, whose purpose is to control production equipment and/or process parameters.
QA	Quality Assurance: (1) Verification, by inspection and audit, that equipment and the procedures according to which it is operated result in products consistently of the required *quality (1)*. (2) The department within a company responsible for such inspections and audits.
Quality	(1) Specification of measurable criteria that have been defined to ensure acceptability of a product. (2) Conformance to such criteria of a product or the process used to prepare it.
Regulations	Requirements published by the Health and Safety Executive (HSE) that, under the *HSW Act*, have the force of law (UK).
Respirable dust	Airborne material that is capable of penetrating to the gas exchange region of the lung.
Risk	A measure of both the severity of a *hazard* and the likelihood that it will result in actual harm to people or to the environment.
Risk phrase (R-phrase)	A standard phrase giving simple information about the *hazards* of a *chemical* in normal use (see Appendix 3).

Robust

(Of a process) able to manufacture products of a consistent *quality (2)* under all foreseeable process conditions.

RPE

Respiratory Protective Equipment, *PPE* specifically designed to protect the wearer from inhaling dangerous concentrations of *chemicals*. It ranges from simple filters over the mouth and nose to self-contained breathing apparatus.

RTP

Rapid Transfer Port, a port assembly used to provide contained transfer directly between an *isolator* and a container. The port doors on such an assembly can open only if it is docked with another matching assembly.

Safety phrase (S-phrase)

A standard phrase giving advice on safety precautions that may be appropriate when using the *chemical* (see Appendix 3).

Sampling train

A combination of sampling head, including a suitable filter medium, and a pump that draws air through the filter; used to measure airborne contamination.

SBV

Split Butterfly Valve, a valve assembly used to provide contained transfer directly between vessels. Each vessel is closed by a disc, which acts as half of a *butterfly valve* and which can rotate about its diametrical axis only if it is docked with another matching disc.

Scheduled process

Processes referred to in HSE Regulations or other Government legislation because of the special hazards they present. Lead manufacture is one example.

Short-term exposure

The airborne concentration of a *chemical* in a person's breathing zone, calculated as a *TWA* over 15 minutes.

SOP

Standard Operating Procedure, detailing in sequence the actions to be carried out in order to operate equipment or carry out any other duty at work.

Step-over

The dividing feature (typically a bench in a changing room adjoining a room where hazardous materials are processed) at which an operator changes from external clothing into clean or other protective work clothes. Having made the change, the operator must not cross the feature again but must leave the hazardous area via a suitable decontamination area.

Source code

The detailed steps carried out by a software process, written in the appropriate programming language.

Substance

A chemical element or one of its compounds, including any impurities.

Substance hazardous to health	A *substance* that can harm human health if it is inhaled or ingested or if it comes into contact with the skin. A more formal definition is given in the *COSHH Regulations* (UK).
Suitable and sufficient assessment	An assessment of the *risks* to health arising from *substances hazardous to health*, the practicability of the prevention of exposure to those *substances*, the identification of the steps that need to be taken to achieve adequate control of exposure where prevention is not reasonably practicable and any other action needed to prevent harm arising from the presence of such *substances*.
Supply law	Legal duties placed on manufacturers and suppliers of new machinery, based on EU Directives and implemented in the UK by the Supply of Machinery (Safety) Regulations 1992.
Supply requirements	Duties of suppliers under the *CHIP Regulations* to *classify hazards* of *chemicals* they supply, inform their customers of those *hazards* and package the *chemicals* safely (UK).
Tactile danger warning	Normally a small raised triangle intended to alert the blind and visually impaired to the fact that they are handling a *substance hazardous to health*.
Technical area	The area of a facility in which production support and ancillary equipment are located or in which products are never exposed, which can be designed to lower environmental standards than process areas.
Toxic	(Of a *substance*) able to damage human tissue or organs so as to cause illness or death. For the purposes of this guide the term is generally used to cover *potent* as well. Appendix A of the US standard 29 CFR 1910.1200 distinguishes between *substances* that are 'toxic' and 'highly toxic' with reference to the doses or concentrations that are lethal to specified laboratory animals. In the UK, *CHIP* distinguishes between 'toxic' and 'very toxic'.
TWA	Time-Weighted Average, an average value of a parameter calculated by summing multiples of all values of the parameter by the duration for which each value applies and dividing the result by the total duration. In the context of this guide, the parameter is normally airborne concentration of a *substance* and the total duration normally 8 hours (long-term) or 15 minutes (short-term).
URS	User Requirement Specification — a formal statement of the required scope and duties of a plant, facility or system.
User law	Legal duties placed on users of machinery and other equipment, based on EU Directives and implemented in the UK by the Provision and Use of Work Equipment Regulations 1992.

Validation	Confirmation by examination and provision of objective evidence that the particular requirement for a specific intended use can be consistently fulfilled.
VMP	Validation Master Plan, the document that details how the *validation* of a process is to be carried out.
Validation review report	A detailed report produced following review of completed *DQ*, *IQ*, *OQ* and *PQ* reports, which states how the protocols for a particular system were implemented and what results were obtained and recommends whether or not the system should be accepted as meeting the requirements of the protocols.
Zone	See *HAC*.
Zoning	See *HAC*.

Appendix 2 – References

This appendix lists the documents referenced in this guide under the following headings:

The number to the right of each entry indicates the chapter or appendix of this guide in which the publication is referenced.

European Directives

Council Directive 89/392/EEC of 14 June 1989 on the approximation of the laws of the Member States relating to machinery. Codified version CF 398L0037. OJL 183, 29.6.1989, p9.   2

Directive 98/37/EC of the European Parliament and of the Council of 22 June 1998 on the approximation of the laws of the Member States relating to machinery. OJ L 1998, 22.7.1998, p1.   2

UK legislation

Control of Asbestos at Work Regulations 1987.   2

Control of Lead at Work Regulations 1980 (S.I. No. 1248).   2

Control of Substances Hazardous to Health Regulations 1999 (S.I. No. 1999/347).   2

Chemicals (Hazard Information and Packaging for Supply) Regulations 1994 (S.I. No. 1994/3247, amended by S.I. 1996/1092, 1997/1460, 1998/3106, 1999/197).   2, A3

Health and Safety at Work etc. Act 1974.   2, 5

Health and Safety (Dangerous Pathogens) Regulations 1981.   2

Health and Safety (Miscellaneous Modifications) Regulations 1993 (S.I. No. 745).   2

Mines and Quarries Act 1954.   2

Personal Protective Equipment at Work Regulations 1992 (S.I. No. 2966).   2

Personal Protective Equipment (EC Directive) Regulations 1992. (S.I. No. 3139).   2

HSE publications

US legislation

Table A3.1 (*Continued*)

Code	R-phrase
R53	May cause long-term adverse effects in the aquatic environment.
R54	Toxic to flora.
R55	Toxic to fauna.
R56	Toxic to soil organisms.
R57	Toxic to bees.
R58	May cause long-term adverse effects in the environment.
R59	Dangerous for the ozone layer.
R60	May impair fertility.
R61	May cause harm to the unborn child.
R62	Possible risk of impaired fertility.
R63	Possible risk of harm to the unborn child.
R64	May cause harm to breast-fed babies.
R65	Harmful: may cause lung damage if swallowed.
R66	Repeated exposure may cause skin dryness or cracking.
R67	Vapours may cause drowsiness and dizziness.

Table A3.2 Basic S-phrases (from L100)

Code	S-phrase
S1	Keep locked up.
S2	Keep out of the reach of children.
S3	Keep in a cool place.
S4	Keep away from living quarters.
S5	Keep contents under (*appropriate liquid to be specified by the manufacturer*).
S6	Keep under (*inert gas to be specified by the manufacturer*).
S7	Keep container tightly closed.
S8	Keep container dry.
S9	Keep container in a well ventilated place.
S12	Do not keep the container sealed.
S13	Keep away from food, drink and animal feeding stuffs.
S14	Keep away from (*incompatible materials to be indicated by the manufacturer*).
S15	Keep away from heat.
S16	Keep away from sources of ignition — No smoking.
S17	Keep away from combustible material.
S18	Handle and open container with care.
S20	When using do not eat or drink
S21	When using do not smoke.
S22	Do not breathe dust.
S23	Do not breathe gas/fumes/vapour/spray (*appropriate wording to be specified by the manufacturer*).
S24	Avoid contact with skin.
S25	Avoid contact with eyes.
S26	In case of contact with eyes, rinse immediately with plenty of water and seek medical advice.
S27	Take off immediately all contaminated clothing.
S28	After contact with skin, wash immediately with plenty of (*to be specified by the manufacturer*).
S29	Do not empty into drains.
S30	Never add water to this product.
S33	Take precautionary measures against static discharges.

(continued)

Table A3.2 (*Continued*)

Code	S-phrase
S35	This material and its container must be disposed of in a safe way.
S36	Wear suitable protective clothing.
S37	Wear suitable gloves.
S38	In case of insufficient ventilation wear suitable respiratory equipment.
S39	Wear eye/face protection.
S40	To clean the floor and all objects contaminated by the material use (*to be specified by the manufacturer*).
S41	In case of fire and/or explosion do not breathe fumes.
S42	During fumigation/spraying wear suitable respiratory equipment (*appropriate wording to be specified by the manufacturer*).
S43	In case of fire, use (*indicate in the space the precise type of fire-fighting equipment. If water increased the risk add:* **Never use water**).
S45	In case of accident or if you feel unwell seek medical advice immediately (show the label where possible).
S46	If swallowed, seek medical advice immediately and show this container or label.
S47	Keep at temperature not exceeding °C (*to be specified by the manufacturer*).
S48	Keep wetted with (*appropriate material to be specified by the manufacturer*).
S49	Keep only in the original container.
S50	Do not mix with (*to be specified by the manufacturer*).
S51	Use only in well-ventilated areas.
S52	Not recommended for interior use on large surface areas.
S53	Avoid exposure — Obtain special instructions before use.
S56	Dispose of this material and its container to hazardous or special waste collection point.
S57	Use appropriate containment to avoid environmental contamination.
S59	Refer to manufacturer/supplier for information on recovery/recycling.
S60	This material and/or its container must be disposed of as hazardous waste.
S61	Avoid release to the environment — Refer to special instructions/safety data sheet.
S62	If swallowed, do not induce vomiting: seek medical advice immediately and show the container or label.
S63	In case of accident by inhalation: remove casualty to fresh air and keep at rest.
S64	If swallowed, rinse mouth with water (only if the person is conscious).

Index